JIA ZHANGKE
SPEAKS OUT

Texts By and Interviews With Jia Zhangke

*Translated by Claire Huot, Tony Rayns,
Alice Shih, and Sebastian Veg*

Jia Zhangke Speaks Out: The Chinese Director's Texts on Film
Copyright © 2015
By Jia Zhangke
Translated by Claire Huot, Tony Rayns, Alice Shih, and Sebastian Veg.

Distributed by Transaction Publishers
10 Corporate Place South, Suite 102
Piscataway, NJ 08854

All rights reserved. Exclusive English language rights are licensed to Bridge21 Publications, LLC. No part of this book may be used or reproduced in any matter whatsoever without written permission from the publisher except in the case of brief quotations embodied in critical articles and reviews.

本作品原由北京大学出版社于2009年出版。
英文翻译版经北京大学出版社授权于全球市场独家出版发行。

The Chinese edition is originally published by Peking University Press in 2009. This translation is published by arrangement with Peking University Press, Beijing, China.

Cover photograph by Gareth Cattermole, Copyright © Getty Images
Cover Design by Bentsion Janashvili
Copy Editing by Theory Editorial
Page Design and Layout by Lorie DeWorken, Mind*the*Margins, LLC

ISBN 978-1-62643-028-0 Paperback

CONTENTS

Building No. 7: Jia Zhangke at Work
    Jia Zhangke Interviewed by Claire Huot in 2014 1

Jia Zhangke: A Different Kind of Animal
    2008 Preface by Chen Danqing. 19

PART I: The Director's Notes on His Films 27

    1. *Xiao Shan Goes Home*
        小山回家 *(Xiao Shan Hui Jia)*, 1996 29

    2. *Xiao Wu*
        小武 *(A.K.A The Pickpocket)*, 1998. 35

    3. *Platform*
        站台 *(Zhantai)*, 2000 41

    4. *In Public*
        公共场所 *(Gonggong Changsuo)*, 2001. 49

    5. *In Public* in My Own Words 51

    6. *Unknown Pleasures*
        任逍遥 *(Ren Xiao Yao)*, 2002 57

    7. *The World*
        世界 *(Shijie)*, 2004. 59

    8. *Still Life*
        三峡好人 *(Sanxia Haoren)*, 2006 65

    9. *Dong*
        东, 2006. 67

10. Useless
   无用 *(Wuyong)*, 2007............................. 69

11. *24 City*
   二十四城记 *(Ershi Si Chengji)*, 2008.............. 71

PART II: The Making of a "Grassroots" Director 73

12. I Don't Romanticize My Experiences................ 75

13. I Have More Headaches than the Monkey King....... 79

14. When Only Wine Can Unleash
    Your Stream of Consciousness...................... 85

15. A People's Director from the Grassroots of China
    A Conversation between Lin Xudong and Jia Zhangke.. 89

16. Images Taken from the World of Experience
    *Interview by Correspondence between
    Sun Jianmin and Jia Zhangke*...................... 123

17. The Genetic Composition of My Films 151

18. Make Films According to Your Own Beliefs
    *A Conversation between
    Hou Hsiao-hsien and Jia Zhangke*.................. 153

19. It's a Narrative as well as a Documentary
    *A Conversation between
    Tsai Ming-liang and Jia Zhangke*.................. 161

PART III: The Director as Film and Cultural Critic 175

20. Summer in Tokyo (1999)........................... 177

21. Who is Ushering Chinese-Language
    Cinema into the New Century?.................... 183

22. The Unstoppable Movement of Images:
    New Films in China since 1995 189

23. I Heard the Spring of Cinema is Upon Us 205

24. The World Sits on a Tatami 209

25. We Need to Recognize the Defect in our Genes
    (*A Lecture*) 217

26. Martin Scorsese, my "Elder" 223

27. Everyone's Body is a Work of Art in Itself
    *A Conversation between Liu Xiaodong and Jia Zhangke
    Conducted by Deng Xin and Wang Nan* 229

28. Recognizing the Beauty of One's Own Body
    *A Conversation between
    Tony Rayns and Jia Zhangke* 243

29. On *Useless*
    *A Conversation between
    Tony Rayns and Jia Zhangke* 247

PART IV: Thoughts of a "Folk Director" 255

30. The Age of Amateur Cinema is About to Return (2001) .. 257

31. Now That We Have VCDs and Digital Cameras (2001) .. 261

32. Have You Bought Jia Kezhang's *Platform*? 267

33. Letter To The Yamagata International
    Documentary Film Festival 271

34. Fireworks Thundering but VCR Blundering 273

35. This Year Will Eventually Come to an End 277

36. Darkness and Light Of 2006 281

37. Bewilderment.................................. 285

38. The Cowardice of Our Entire Generation
    A Lecture at Peking University.................... 289

39. Blockbusters are like a Contagion and Destroy Social Values
    A Conversation between Xu Baike and Jia Zhangke ... 297

40. Deciphering China through the Eyes of a Film-Poet
    *A Conversation with Dudley Andrew, Ouyang Jianghe,
    Zhai Yongming, Lu Xinyu, and Jia Zhangke* 319

Index of Names..................................... 333

Index of Films & Media............................. 337

Index—Other 341

BUILDING NO. 7: JIA ZHANGKE AT WORK

Jia Zhangke Interviewed by Claire Huot on May 2, 2014, at 4 p.m., in Beijing

It was a good thing I'd left early for my 4 p.m. appointment with Jia Zhangke at his office. It was not actually on the street he had named but in a residential compound in one of those large Beijing blocks behind that street. Once I'd located the compound, I searched for Building 7. I was relieved to see that the many high rises in the compound were all identified with large, readable plaques. Unfortunately, I could not find number 7. Some of the buildings were marked with numbers, others with letters. There seemed to be no order or continuity to the numbering. I was beginning to regret my request to meet in his workplace rather than in some café, restaurant, or hotel bar, as is so often the case in China. A little nervous, I asked a passing resident for Building 7 in Chinese. She pointed me in the direction of Building G. In the lobby, the receptionist nodded without looking up from his smartphone. "Yes," he said, "Building 7 right here." "Here? But it's not written 7," I said. "The sign above the door is G." He looked up at me with an expression that seemed to say, "How can a foreigner be smart enough to speak Mandarin and yet too dumb to read an address?" He counted on the fingers of his right hand: "A, B, C, D, E, F, G." Of course, I realized, G is the seventh letter of the Western alphabet. He smiled. Everyone knows that.

The building is one of those nineties residential high rises that still echoes the socialist block houses, but with a cleaner, better quality and an elevator that works. The door to Jia Zhangke's office

was slightly open. I knocked and from the inside he called out, "Please come in!"

Jia came out of the kitchen where he had been preparing tea. He wore a navy blue T-shirt with a big white star on it, dark jeans, worn-out sneakers, and an inexpensive 1950s Chinese watch. He looked tired but gave me a welcoming smile. "Twenty years," he said. I said, "Twenty years." We sounded like the two protagonists in Fei Mu's 1948 film Spring in a Small Town, *only in their case the meeting takes place ten years later*. Jia Zhangke and I hadn't met since I interviewed him when he was still a student at the Beijing Film Academy. He had just screened his film Xiao Shan (Going Home) to an audience of migrant workers, and I'd read about it in the Beijing Evening News. Unlike the other artists I was interviewing then, Jia had come to the meeting with his team, the Youth Experimental Film Group, which included Wang Hongwei, who would later become famous as the protagonist in Jia's Xiao Wu, and his sound engineer Lin Xiaoling, a woman, which caught my attention because I had met so few female artists during that time in China.

The two main rooms of the apartment serve as Jia's office. The walls of both rooms are decorated with large framed posters of his films. Separating the two rooms are two wide glass bookcases with several shelves displaying his numerous awards, including the most recent: the 2013 Cannes Film Festival Best Screenplay Award for his film A Touch of Sin.

In the front room, as I deposited my handbag, I noticed a large table to the left that contained piles of books about Jia Zhangke and his works in Chinese, English, French, and probably other languages. The most prominent book was propped up atop the pile: his own writings, Jia Xiang, the original Chinese version of the book you are now reading. On the wall, beside the entrance, was a Sony *screen in front of four or five comfortable chairs, a perfect setup for private viewing.*

Jia Zhangke led me past the trophies into the next room by the open window. The day was gorgeous. I detected a faint trace of cigar smoke. Jia sat by his desk, on which lay a cheap Chinese laptop. Behind him was the poster for 24 City *and a life-size picture of Zhao Tao, who has been his favorite actress in all his films and, since 2012, his wife. On a table against the other wall was an iMac computer and a second monitor, as well as two teetering piles of books. Against the wall behind me was a large, long, glass-cased bookshelf containing hundreds of DVDs; I noticed that Jia Zhangke had all of Jim Jarmusch's films. Jia Zhangke rested an elbow on the table, linked his two hands, and looked at me. I turned on the recorder. Jia Zhangke maintained this posture and acute attention throughout the interview, except when he took sips of tea from his thermos.*

Claire Huot: Your book *Jia Xiang* contains several wonderful stories about you becoming a filmmaker. Some are quite humorous: your meeting with Martin Scorsese, your discovery of a pirated copy of your film *Platform*. Some stories help us understand moments in your films: for instance the fact that there was no train in your hometown so you would pedal to the next town to see and hear trains, just as the characters in *Platform* do. Other stories help us understand where your respect and admiration for country folk comes from: your illiterate "Nanny" taught you how to behave as a human being. You are a great storyteller. Since the publication of your book in 2009, would you like to tell us of an event, person, or a place that's had a decisive impact on you?

Jia Zhangke: First of all, since publishing this book in 2009, I have continued to write and I'm hoping to publish its sequel with the same title, *Jia Xiang 2*. Writing is very important to me. It's a way I get to understand myself and organize my thinking process. Even though I'm very busy, I still find time to write. And it's also linked to my film work. There have been quite a few important changes to the latter in the past few years. One very important thing is the making of *A Touch of Sin*, which was released in 2013. Before that, from 2008 to 2010, my films *24 City*, and *I Wish I Knew* were historical, about the past. So are my earlier films about my hometown,[1] which are also about history, the local history of my region, and about the recent past. I did these because I'm interested in investigating present-day Chinese society after all of our political reforms. My first films were set in the years I was growing up, which happened to be the beginning of the Reform Era in the early 1980s. Another period that is crucial to understand is the very beginning of modernization in Chinese society. I've studied China's history during the late Qing dynasty, from 1895 to 1905, when there were the first signs of modernization. The Boxer Rebellion (the Militia United in Righteousness), a pro-nationalist movement has not yet been discussed in depth in China. We still understand it from the point of view of the Communist Party. That movement is crucial for rethinking the origins of modern Chinese society. That was also the time when the imperial examination

---

1. The original Chinese title of the book, *Jia Xiang*, means "Jia Zhangke's thoughts," but it calls forth its more familiar homonym "hometown."

system, which had been the traditional form of education for centuries, was abolished. So I read a lot of local histories and was planning to make a film set in old China during that pivotal period.

But then, as of 2009, I became involved in social networks, in micro-blogging. I started following Sina Weibo (a hybrid of Twitter and Facebook), which launched in 2009. Until then, I'd used the Internet simply to write and receive emails. This new practice drew my attention back to present-day China. Every "weibo" feed is a media in itself. I realized that a variety of voices from different regions were continuously providing information that was unavailable on traditional media. They were reporting on the spot as events were happening. In many ways, they were breaking through the censorship around news. Reading blogs became a critical activity in order to understand what was happening. In the past few years, I've read numerous reports of violent incidents that were totally unknown to me previously. I'd never been aware of such things, they were not familiar to me. Blogs brought me back to the "now." So the two main things that have happened to me are my involvement with micro-blogging and my return to present-day situations. *A Touch of Sin* is based on four real-life cases that occurred recently in China and that I discovered as a follower of Weibo.

Huot: In several of the texts, and especially in the essay "*'In Public' in My Own Words*," you describe your attitude when you're on a shooting location. First you stand there and absorb the atmosphere, the sounds and sights. And you do this for as long as it takes. Are you as patient today as in the past? Do you take the time to feel and experience that space? I ask this question because that way of working is admirable, but I wonder if you have as much free time now as you did in the past. Especially since your most recent films—not mentioned in this book simply because they had not been made (*I Wish I Knew* [*Haishang chuanqi*, 2010] and *A Touch of Sin* [*Tian zhuding*, 2013])—are so complex in terms of characters and cinematic techniques, and they are set in different parts of China. They were probably quite expensive to make. Has your working method changed under these pressures? Your French editor, Matthieu Laclau, has written that in the case of *A Touch of Sin*, you were extremely pressed for time. You would go out to shoot on location and rush back to the studio to edit because you wanted the film to be ready for the Cannes Film Festival.[2]

Jia: Actually, my way of working has not changed much. Even though it's true that I'm doing a great many more things than I did in the past. Every day, I have more or less the same routine. In the mornings, I work on my correspondence and meet people here in my office. After lunch and until sunset is my time to do creative work—I write or edit film. In the evening, I often have to meet more people or attend some social event. Plus, my work as a producer [for Jia Zhangke's own Xstream Pictures] is more and more demanding. But I set an annual

---

[2]. "Témoignage de Matthieu Laclau, monteur du film *A touch of sin*," [Account by Matthieu Laclau, film editor of *A Touch of Sin*] *Chine et films*, December 5, 2013. http://www.chine-et-films.com/article/temoignage-de-matthieu-laclau-monteur-du-film-a-touch-of-sin.

agenda to follow. For example, this year, from January to May 2014, I'm mainly promoting my latest film, *A Touch of Sin*. I've just returned from the USA [April 30] and am soon leaving for London, where there's an event on May 16. But, as of June, the other half of the year is my time to create, to work on a new film.

Of course, not every year is exactly the same. For example, *A Touch of Sin* was made very quickly compared to my other films. Both the writing of the script and the filming were done in a few months—five to six months. But there was a lot of thinking and research prior to that. I began collecting material three years earlier. In those three years, I found a dozen or so real-life cases, but it took me some time to decide which ones to use and, especially, how I would film them. The most important and longest process was deciding on the structure I would use to make *A Touch of Sin*. I couldn't find an approach, the right film language to do the film. But in the fall of 2012, I suddenly found it! I realized that these events strongly recalled those in the thirteenth-century Yuan dynasty novel, *Water Margin* (*Shui Hu Zhua*).[3] The stories in the real-life cases and the characters were quite similar to those in that traditional martial arts novel from so many centuries ago. They are larger than life characters who are avengers living outside of the system. I could use the narrative style of the novel that follows one character until something happens to him or her and then drops that character and moves on to another story. I could depict each

---

3. The classic novel is also known as *Outlaws of the Marsh* and tells the stories of convicts, outlaws, and marginal characters of society.

character's fight against corrupt people by filming their actions as if they were engaging in martial arts. I wrote the scenario feverishly and rushed to shoot the film. But before that, there was a very long conception process.

Huot: This book also includes talks on cinema that you delivered to students and professors at the Beijing Film Academy in which you decry the lack of knowledge of pre-1949 Chinese films. You analyzed the way the 1937 film *Street Angel* (*Malu tianshi*) by Yuan Muzhi was ingeniously constructed and shows a Chinese urban community life that is no longer to be found in post-1949 films. You also often refer to *Spring in a Small Town* (*Xiao cheng zhi chun*, 1948) by Fei Mu as another jewel of invention at a time when filmmakers had no examples to build from. It's as if you're saying that Chinese students of Chinese cinema should not rely solely on canonical views coming from Taiwan or Hong Kong, or from the Chinese Mainland film establishment. You argue that the renowned Yasujirō Ozu, whom you admire greatly, is not, however, the only important Asian filmmaker. China had its own Golden Age, its own great masters.

By providing analyses of these pre-1949 films and showing how they have inspired you, do you think you are encouraging young people to look to the past, to explore their own cultural heritage? How great is your influence on the next generation?

Jia: I hope that through my work and my talks I can spread knowledge of China's history and film heritage, especially among the young generation so that they can become more engaged. I find that, unlike in the case of literature, there's a lack of knowledge of cinematic history. In literature, there's a systematic knowledge of works; for example, we know a lot about pre-1949 writers such as Shen Congwen.[4]

Huot: And Zhang Ailing.[5]

Jia: Right, Zhang Ailing. And going further back in time, we study Yuan dynasty plays, Ming dynasty novels, and so on. But in the case of cinema, because of the way we consume it (and the costs associated with it), there has been no systematic study. Sure, occasionally there will be a retrospective, or a few old films will be shown, but without any perspective. It's very hard to get young people to see them. People in the film world have seen them, but there's never been wider, comprehensive screenings for the general public, in particular for the young public. That's very sad.

Huot: Because you, Jia Zhangke, a famous young director, advocate the study of China's old films, will young people want to follow your lead? How influential are you among young people?

Jia: Now people are talking more about these films, there are more retrospectives, books are being published,

---

4. Shen Congwen (1902–1988) was a writer famous for his light, pastoral works written in the 1930s.

5. Zhang Ailing, well known as Eileen Chang (1920–1995) was a writer who is especially famous for her depictions of 1940s life in Shanghai. Her works were not known in Mainland China until the 1980s.

and they're discussed on the Internet. So it's changing. Compared to when I gave those talks, things have improved quite a bit.

Huot: In your book, there's one particular passage where you discuss *Street Angel* and start by saying that, of course, this film is a left-wing film, but having said this, there's a lot more to it.

Jia: These days, if you say a film is a "left-wing" film, young people roll their eyes, and say, "Boring." So, I say, it doesn't matter whether a film is "left-wing" or "right-wing"; as long as it's a good film, it's worth studying. We should not pay heed to such categories or concepts. And then, there are great silent films, such as Wu Yonggang's *The Goddess* (*Shennü*, 1934). Young people are also uninterested in silent films. Lately, in the past few years, I've been promoting silent films and their charm. I'm not saying a film like *The Goddess* is great, but silent films are the beginning of cinema; they have their cinematic language and valuable, noteworthy moments. They need to be studied, too. Once I present the importance of silent films in cinematic history, I then talk about specific films such as *The Goddess*. People get it.

Huot: I read somewhere that you were affiliated with a college. Is it the Xi'an Film College?

Jia: No, I'm with the Academy of Fine Arts here in Beijing. But I don't teach, I have graduate students. Right now I supervise three students. I especially maintain ties with the young generation as a producer. I'm the producer of several films made by young people. For instance, *Seeking Advice* (*Wen dao*), a recent documentary that consists of interviews of twelve important figures in Chinese culture and society.[6] There are activists in environmental protection and for the prevention and cure of AIDS, and also entrepreneurs and artists like Xu Bing. The question they were asked was: "What was the most difficult time for you, and how did you surmount it?" These cases are meant to encourage young people to do their own thing and not fear difficulties.

Huot: I read the interview with Xu Bing, China's most renowned artist, both inside and outside of China. He described his rough times during the Cultural Revolution and also while living in New York. I noticed the interviewer, Wei Tie, called him "xiansheng" (like the Japanese honorific "sensei"). Is this a new trend in China?

Jia: Oh, I also call Xu Bing, "xiansheng."

Huot: Why? Because he's older? After all, you're both equally renowned artists...[7]

Jia: Because he inspired me a lot. Especially in the late 1980s and 1990s. His spirit of rebellion. An anti-bourgeois

---

6. Jia Zhangke wrote the script for this film, which was published under that same title in 2011.

7. When I mentioned this to Xu Bing several days later, Xu Bing said: "Oh, but Jia Zhangke is much more famous!"

campaign at the time criticized Xu Bing. That's how I got to know his work.

Huot: Which year?

Jia: I've forgotten, but it must have been in the early 1990s. I remember I had not yet started university. They were criticizing his *Book from the Sky*.[8] I read the article condemning him and I marveled at his creativity and his courage. Xu Bing inspired a whole bunch of emerging artists like me.

Huot: Outside China, you have a great reception. Every major city shows your films. Critics abroad, without exception, acclaim your works. You are considered China's most important film director. On the flip side, it seems that your films don't make many appearances within China. For example, your 2013 film *A Touch of Sin* has not been banned but is not being shown in China. Is that correct?

Jia: [Jia laughs sadly.] You've put it exactly right.

Huot: And your former films, *Platform* (2002) and *Still Life* (2006) were shown in second-tier theaters. They were eclipsed by blockbuster productions. How do you feel about this dichotomy? That is, an outstanding reception abroad and almost nothing at home?

---

8. *Book from the Sky* is a monumental work of woodblock printed characters that look like Chinese characters but are Xu's inventions.

Jia: Up to now, three of my films cannot be shown in China: my first feature-length films, the trilogy, *Xiao Wu*, *Platform*, and *Unknown Pleasures*. As for my latest film, *A Touch of Sin*, it passed censorship last year (2013), but so far we have not been allowed to distribute it. It's a very strange new phenomenon. But it's a fact. As for 2006, when *Still Life* was released, it came out at the same time as Zhang Yimou's *Curse of the Golden Flower* (*Man cheng jin dai huang jin jia*), so it found itself in a very bad time frame. To a certain degree, the Chinese government is still controlling film and what the public can view. That goes for *A Touch of Sin*. For a while, the Chinese media were not even allowed to report or discuss it. Which is terrible because a film needs to be introduced and talked about. But on the other hand, this is the age of the Internet and those who love this kind of film are discussing it online.

Huot: Are they downloading the film?

Jia: There are already pirated versions on the Net. That started as early as March 2014. This is a serious blow against our distribution of the film. It's a real predicament: we can't show the film publicly, but people are watching it.

But this does not affect my creative work all that much. What's happening with *A Touch of Sin* has to do with the present political climate and the cultural climate today in China. I'm constantly reminding myself to shut it all off—do my work and forget such things.

I have to be free, free from all of this. I remember telling myself, when I was filming *Platform* in 2000 or so, that I would not compromise to pass censorship; nor would I go to extremes in the film just because it wasn't going to pass censorship anyway. You have to work with peace of mind and follow what you truly intend to do, without paying heed to what's happening around you.

Huot: Your latest film, *A Touch of Sin*, draws parallels between the fate of humans and animals. In two earlier short works, *The Condition of Dogs* (*Gou de zhuangkuang*) and *Black Breakfast* (*Heise zaocan*),⁹ you expressed concern for animal welfare and the protection of the environment. It seems to me that this is very different from your other films, which are human-centered, about individuals' personal experiences, their lives in the context of China's socio-political changes. I wonder if your worldview is shifting toward a post-humanist stance. Are you now more interested in depicting humans in their connection with other animals, and in their natural environment?

Jia: This is a recent change in my way of thinking, which has taken place only in the past few years. It stems from my own personal experience. I'm now considering humans in their physical environment and am concerned about ecology. I come from Shanxi Province, and Shanxi is rich in coal. There are a lot of factories

---

9. *Black Breakfast* is a segment from the international omnibus film *Stories on Human Rights*, and it focuses on environmental concerns.

producing electricity and other derivatives that are chemically engineered from coal. And that's seriously ruining the environment. Also, in 2006, my father, who was a very healthy man, was suddenly diagnosed with lung cancer. He died very quickly thereafter. When he was getting treatment in a Beijing hospital, I realized that many of the patients with that condition were from Shanxi. That's when I really understood the disastrous impact of Shanxi's environment on individuals' lives. From that time on, I started to be aware of the natural environment, because humans are part of it. As for *The Condition of Dogs*, that was in 2001 when I went to a market that sold dogs.

Huot: In Beijing?

Jia: No, in Datong. I made the documentary on the spot. It was fortuitous. I was moved by the courage of one of the puppies that was stuffed in a sack. That little one, in spite of his impossible situation, gradually managed to bite and claw away until he could poke his head out of the sack. I filmed him.

As for A Touch of Sin, the film does represent a change in my way of investigating life. It's true that we humans all look at life from a human point of view, but if you look from another perspective—for example, from the animals' perspective—it's relatively easy to understand the human condition. So for this film I consciously changed and sought to use a non-human perspective to depict reality. So yes, my perspective on the world, my mode of investigation, has been transformed.

Huot: One final question. You have generously accepted to be interviewed countless times. These interviews are usually to promote a new film, before and after its release. Today's interview serves as an introduction to your book now translated in English. Do you have anything in particular you'd like to say to your English readers that you've never had the opportunity to say, or a statement you'd like to reiterate?

Jia: I find that writing is a supplement to filmmaking. Because a film is a long, drawn-out process. It takes me two years to make a film. Half a year to write the script, the other half to film it. Then there's all the time spent to promote and discuss it. A film takes a long time from inception to reception. But an idea can come in a flash and often does. So I take the time to write. Writing allows me to anchor myself and explore those ideas that come to me, more or less as they come to me.

Apart from my life as a filmmaker, I have a life as a writer, which is also indispensable and which I thoroughly enjoy. Aside from that, there's nothing else.

Huot: Oh, I wanted to congratulate you and Zhao Tao on your marriage. Wishing you a lifetime of happiness.

Jia: We will!

Huot: You're always so busy. Does Zhao Tao travel with you?

Jia: Oh yes. Well, sometimes. She came with me to the Toronto International Film Festival last year.

I walked out of the apartment office. Jia Zhangke thanked me and said, "I hope you come back to China often. Although the air gets worse and worse." As the elevator bell rang he added, "When you did that first interview, twenty years ago, you arrived by bicycle." What an amazing memory, going back to one of the countless people he's met over the past twenty years.

It occurred to me at that moment that the old 1950s watch Jia Zhangke was wearing must have been his father's.

JIA ZHANGKE: A DIFFERENT KIND OF ANIMAL

*2008 Preface
By Chen Danqing[1]*

Translated by Claire Huot

Today Jia Zhangke screened his film *Xiao Wu* (AKA *The Pickpocket*) here. How time flies....

I have an old friend from Shanghai, Lin Xudong—we met when we were seventeen or eighteen years old. We grew up together, did some oil painting together, and were both sent to the countryside in Jiangxi. We went our separate ways in the eighties. He stayed in China and I went to New York, but we kept up a correspondence. We've been friends for almost forty years.

As students of Western oil painting, we wanted desperately to see the original works, which is why I left China. Lin, being less hasty, stayed behind. He realized that in film, there's no such thing as the original work. "I can watch *The Godfather* just as well here in Beijing as in Rome," he told me, "it's the same movie." So he studied film and ended up knowing everything about the world of cinema—the contending schools and trends and aesthetics. When he graduated from the Central Academy of Fine Arts, he was assigned to the Broadcasting Institute to teach film history.

In 1998, he suddenly made an overseas phone call to me. "Someone named Jia Zhangke has just come out with a film called *Xiao Wu*," he told me. There had to be a very good reason for Lin Xudong to make that call.

---

1. Chen Danqing, b. 1953, is a realist painter whose first series of oil portraits of Tibetans propelled him to fame in the early 1980s. His talk served as the preface to Jia's book in Chinese.

Few people have actually met the very gifted Lin Xudong. And yet, he's been involved in many important events and projects in China. These include his many contributions to underground films and documentaries, where he's the hero behind the scenes. Twice, Lin single-handedly organized international symposia in Beijing on documentary cinema, to which he invited a number of important documentary filmmakers from Europe and America.

Since the eighties, Lin Xudong had been following the rise of the so-called Fifth Generation of Chinese filmmakers. He later became acquainted with Sixth Generation directors such as Zhang Yuan and Wang Xiaoshuai. During his correspondence with me throughout the nineties, he talked about the various developments in Chinese cinema. Initially he was very excited by the films of the Fifth and Sixth Generations, but he became gradually disappointed. By the end of the nineties, the Fifth Generation directors had made their best films; they were in a slump, they had ceased to create masterpieces. The Sixth Generation directors, after their first outpouring of movies, had produced no significant works either. Then, in the middle of the night, a very serious Lin Xudong phoned me to say that he was express-mailing me a video copy of *Xiao Wu*. I received it shortly after, and, on viewing it, I understood why he'd phoned me so urgently.

In the spring of that same year I was asked to substitute for a teacher at the Central Academy of Fine Arts. There, I viewed *Xiao Wu* again, this time in the presence of Jia Zhangke himself. This was in the late nineties, when the filmmaker was showing *Xiao Wu* in post-secondary educational institutions. In those days, only 16 mm prints were available; you couldn't add subtitles, and the film was a medley of dialects from Shanxi and the North

East. Consequently, at screenings, Jia Zhangke himself would provide simultaneous translation into standard Mandarin. The hall at the Central Academy of Fine Arts is relatively small. I was seated in the tenth row or so, and Jia Zhangke stood at the very back, under a dim light. Every time a character spoke, he would provide his simultaneous translations. That's how I saw *Xiao Wu* for the second time. It was a very peculiar viewing experience. Later I saw the film again, for a total of three screenings.

In 2000, I returned to China for good, just in time for the completion of Jia Zhangke's second film, *Platform*. In the middle of the night, he called me and Ah Cheng to come over and see the final edit of the movie. It was summer and the streets were so hot that, just walking a few steps, you became sticky with sweat. After returning to China I saw, in succession, his third, fourth, and fifth film—and I've just recently seen *Still Life*. I've been fortunate to follow such a director's entire oeuvre from his directorial debut.

At this point, I'd like to mention another person: Liu Xiaodong is a young teacher at the Central Academy of Fine Arts. He is only a few years older than Jia Zhangke. Jia Zhangke was born in 1970 and Liu Xiaodong in 1963. In 1990, I chanced upon Liu Xiaodong's paintings in a fine-arts periodical I picked up in New York's Chinatown. I became very excited, just like Lin Xudong when he saw Jia Zhangke's *Xiao Wu* in 1998. I thought, "Alright! China has finally produced such a painter." I immediately wrote to him and rapidly received a reply. I learned that we shared the same alma mater. Liu painted his first work in 1988 and has continued to paint ever since. In my opinion, Liu Xiaodong is the Jia Zhangke of the fine arts world, just as Jia Zhangke is the Liu Xiaodong of film.[2]

---

2. Liu Xiaodong, who is a friend of Jia, is the subject of Jia's 2006 documentary *Dong*.

What do I mean?

My generation has repeatedly claimed that we are pursuing realism and humanism and that art must truly express the times. But, in fact, none of us has realized this: the Fifth Generation filmmakers didn't succeed, and neither did I, and the generation before us also failed. In their case, they were not even allowed. National policies did not allow people to tell the truth. My generation's failure is due to the fact that, after a long period of not being allowed to speak the truth, once we were told it's alright to do so, we didn't know how.

After the Great Cultural Revolution, people like me caught the eye of the critics, but in reality this was simply because the scene had been so bleak and desolate before. Some ten years later, Liu Xiaodong flung his ferociously raw paintings at us. Life under his brush looked like a heap of dung. It was so real. His paintings overflowed with passion and youthful vigor. He was twenty-seven or twenty-eight years old when he started creating these works, throwing handfuls of dung. At first, the fine-arts community didn't react. It took them a few years to realize "Hey! This guy's really something else." His first paintings depicted laborers and bored, restless youth with nothing to do and no place of their own under the sun. Several years later, a young filmmaker named Jia Zhangke came up with *Xiao Wu*, the story of a thief, a young loser.

Ten years ago, in New York, when I inserted the *Xiao Wu* tape and witnessed the birth of the character Xiao Wu, I remember saying, "Someone finally got it right!" This young hooligan from the North, smoking a cigarette, twitching a leg, was exactly right. The films of the Fifth Generation contain nothing truthful. Xiao Wu is a little bum from a country town that can be found anywhere in China. At the beginning of the film, he's a youngster with no goal, no position, no future, standing on the roadside waiting for the bus. Subsequently, he bums around for the entire duration of the film until he ends up squatting on the street in handcuffs, encircled by gawkers. From start to finish, it's right on the mark.

China's small towns are full of Xiao Wus. And yet, until then, no one had portrayed them in any medium. But young Jia Zhangke, in his first try, captured him perfectly. When I was in Taiwan this year with Hou Hsiao-hsien, I asked him about Jia Zhangke. Hou Hsiao-hsien said, "When I saw his first film, I noticed that he knew how to work with non-professional actors; knowing how to do that is a sure sign that he's a filmmaker who can do things." And Hou Hsiao-hsien speaks from experience. I find that, compared with Chen Kaige, Zhang Yimou, and Feng Xiaogang, Jia Zhangke is a different kind of animal.[3]

Lin Xudong and I are old "sent-down youths" (*zhiqing*);[4] we never speak about the existential "self." People like Liu Xiaodong shout out their personal anger and frustration; people like Jia Zhangke spell out their generation's despair.

If we look at the wider picture we might say that, after the Second World War, Western cinema was constantly portraying this type of youthful experience. The old culture had vanished and new culture was yet to take its place. Young people were growing up in despair—anxious, confused, not knowing how to act. They knew they were human beings, but they didn't know

---

3. Hou Hsiao-hsien is the most critically acclaimed filmmaker in Taiwan. Chen Kaige and Zhang Yimou are the foremost representatives of the Fifth Generation filmmakers. Feng Xiaogang is the foremost commercially successful filmmaker in China.

4. The *zhiqing* were youth who were forcibly or voluntarily removed from urban environs to rural ones, and converted from students into farmers from late-1950s through Cultural Revolution-era China.

how to act. This condition, which first appeared in the West and then in Japan, became the stuff of screen legends. From the late fifties, a thread emerges stringing together a long list of titles, starting with *The 400 Blows*, *Breathless*, and *Cruel Story of Youth*, all of which use the camera to follow the trail of a young male between adolescence and adulthood, looking at the world through his eyes and his experiences.[5] China joined that thread very late; Chinese artists took a long time to realize: "Oh! This can be told, it can become a painting, a film."

In the eighties, a number of artists, musicians, and filmmakers left China and settled in New York. We were a small circle, and whenever we learned of a new arrival, we'd find a place to eat together and chat. That's how I met Chen Kaige. At the time, I hadn't seen *Yellow Earth*,[6] but the young Chen impressed me; he cut a fine figure and looked like a cool film director. When *Yellow Earth* was shown in New York, I was really eager to see it, but as I sat there in the theater watching, I recognized the movie he'd made was just another film toeing the party line, another story about the Eighth Route Army, folk songs, the Yellow River—the old bag of tricks. At that time in New York, I'd been waiting for *Yellow Earth*, for the Fifth Generation, imagining they would produce authentic films like those of Jia Zhangke. What I saw was a succession of outmoded Japanese-style long takes. I was too uncomfortable to tell Chen then; we were good friends. Only now that decades have gone by, I dare to say so. If I'm exaggerating or being offensive, I'm sorry.

The Fifth Generation filmmakers and I are of the same

---

5. *The 400 Blows* (1959) by François Truffaut, *Breathless* (1960) by Jean-Luc Godard, and *Cruel Story of Youth* (1960) by Nagisa Oshima.

6. *Yellow Earth* (*Huang tudi*), made by Chen Kaige in 1984, was the first Fifth Generation film to be critically acclaimed internationally.

epoch; we were raised on revolutionary films. Around the end of the Great Cultural Revolution, our viewing repertoire was a limited number of Japanese and European films. We fell in love with the long take, admired the color effects of Kodak film, and were won over by the poetic atmosphere and what passed for a philosophical cinematic style. But, although the Fifth Generation abandoned the tenets of proletarian revolutionary cinema and took up the techniques and style of foreign cinema, they did so before they had fully assimilated these.

Director Xie Jin passed away this year.[7] The Fifth Generation has not surpassed the work done by earlier generations. The success of the Fifth Generation is really due to the fact that they were the first directors allowed to participate in international film festivals, to go abroad and collect awards.

Anyone revisiting the films of the Republican era, or those made by New China's First Generation of directors—films like *Storm* (*Fengbao*) (1959), which advocates revolution, or the refined and literary *Early Spring* (*Zao Chun Er Yue*) (1963)—must surely agree these films are all of very high quality. *Early Spring* was made by a left-wing youth who came back from Yan'an. The film is infused with the literati atmosphere of the lower Yangtze. It has a 1930s feel, a good script, and a fluid, unhurried tempo. I don't believe the Fifth Generation has surpassed their predecessors; they've simply been lucky. They had emerged from the Great Cultural Revolution, and they benefited from the cachet of Red China. With the end of the Cultural Revolution, the West was hungry for information about China, of which they understood very little. Left-wing filmmakers and critics in the West were exceptionally welcoming and held the

---

7. Xie Jin (1923–2008) made films during all of the political periods of the People's Republic of China, from the 1950s to the early 2000s.

Fifth Generation in high regard. In fact, at the time, Chinese filmmakers not of that group were largely ignored. This context created the illusion, both in China and the West, that Chinese films were wonderful, mature, and destined to become classics. But that was an illusion.

To speak this way is to offend my own generation, but I am just as unforgiving with myself. I've never forgotten how we set out against a bleak and desolate backdrop. Now that thirty years have elapsed, I expect art to finally authentically express the reality in which we live. No medium can approach the real better than cinema, but in the past thirty years of Chinese cinema, reality and authenticity have been sorely lacking.

I remember Jia Zhangke mentioning in an interview that, as a young loafer in his desolate small town, he'd had many opportunities to drop out, turn bad, and self-destruct. That's a sincere confession. During my days as a "sent-down youth," I also had many opportunities to lose it, to fall apart. But now young people ask: "Who can save us?" My answer may make them uncomfortable: You think like a slave. What is the way to save oneself? Follow your instinct. Do things conscientiously. Don't be impatient. Don't give up. And don't be half-hearted. If you write, every punctuation mark should be noted down properly, no word misspelled. That is how you will save yourself. We must all save ourselves one step at a time. I paint stroke by stroke; Jia Zhangke films frame by frame.

Talk delivered at Peking University Hall, November 23, 2008.

PART I
THE DIRECTOR'S NOTES ON HIS FILMS

XIAO SHAN GOES HOME (XIAO SHAN HUI JIA), 1996

Translated by C. Huot

SYNOPSIS

It's 1995, Chinese New Year in Beijing.

Migrant worker Wang Xiao Shan, who works in the Beijing Hongyuan restaurant, is fired by his boss, Zhao Guoqing. Before heading home, he meets up with people from his hometown, Anyang, who have also come to the capital. Among these are construction workers, a ticket scalper, a university student, a waitress, and a prostitute, but no one is ready to make the journey back home with him. Down on his luck and at a loss, one by one he looks up his old pals staying in Beijing. The movie ends with Wang Xiao Shan stopping at an outdoor barber stall, where he bequeaths his long and tangled city-slicker hair to Beijing.

MY FOCUS

Since I completed *Xiao Shan Goes Home*, I am constantly being asked why I used seven full minutes—a tenth of the film's length—and just two shots to show migrant worker Wang Xiao Shan simply walking. I know that for people in the industry, seven minutes equals twenty-eight commercials, or two MTV videos, and so on. But I'm not interested in that sort of counting. That's the way of measuring things in this business. It's their way. As for me, if an opportunity presents itself to open up a dialogue, I'll use my own way to say what's on my mind.

That's why I let the camera roll and I follow the laborer who's lost his job as he walks the streets around the New Year. The New Year is the time when the old transitions into the new, and we're right there with the camera *and* with the downtrodden Xiao Shan, wandering through a freezing cold Beijing. That long, seven-minute sequence is more than simply gazing. It's *about* gazing, about the capacity to remain focused. Now that our visual and auditory senses have become inured to reshuffling by the second, are there people out there who can join in the unhurried gaze upon what the camera is ultimately presenting? A gaze upon people, some who resemble us, and some who do not?

Constantly switching television channels has transformed our viewing and listening habits. Viewers faced with an incredible variety of audiovisual content tend to make instinctive choices. Artists invariably cater to their tastes, thereby losing their integrity. No one discusses the state of the arts and how to redress it anymore. Artists have come to scoff at art, many of them having adopted the solution of simply abandoning any artistic pretensions. They turn creative work into a procedure. While they may not wish

XIAO SHAN GOES HOME (XIAO SHAN HUI JIA), 1996

to consider themselves pragmatists, they turn art into a practical affair. By staying within professional norms that even stifle passion and initiative, what remains of art other than technique?

If the purpose of such formulaic art is merely to earn a living, well, I would rather be a carefree amateur filmmaker, because I don't want to lose my freedom. When the camera starts rolling, I want to be able to ask myself, "Is everything you see before you really what you're thinking and wanting to express?"

For a while now, the pure expression of moods has become a fad in the arts. Regardless of the medium—visual art, music, or cinema—many practitioners remain on the surface and rarely delve deeper into emotions. In the thousands of shots of the new-generation's MTV-style films, filmmakers ignore life itself and instead focus on the self. The disparate mix of audiovisuals yields nothing but narcissism. Many works are downright onanistic. Their confused perspective fails to reach out to the audience. The artist's vision is no longer clear; it lacks focus. Many people don't have the strength to examine their own feelings because that involves facing up to the imperfections that make us human. Some fast-paced films are devoid of passion; they offer instead an escape from reality. That's why we, the younger generation, once we get behind the camera, should begin by examining ourselves to make sure we're sincere and focused. In *Xiao Shan Goes Home*, the camera does not move all over the place. I wanted to face reality head on, even if reality revealed the usually hidden weak and sordid side of humanity. I wanted to be fully absorbed in each individual scene, absolutely uninterrupted until the next scene. Unlike even Hou Hsiao-hsien who, after shooting a scene in total concentration, lifts the camera toward the distant blue-green landscape to lighten the atmosphere, we choose to maintain our focus. We refuse to back down.

I don't know when I began to feel this way, but certain moments and situations excite me uncontrollably. It can happen at dusk when people flood the streets or at dawn when white steam rises from the outdoor food stalls. I'm suddenly overtaken by a heightened sense of reality. Life, unraveling or taking unexpected twists, flashes before my eyes. Lives come and go, unnoticed. When they pass by, I smell the pungent sweat from their bodies and from mine. Our breaths intertwine, and we've made contact. Individual faces bear similar fates. I prefer to see the pimples on the dirt-covered faces of migrant workers, because their lives are without indulgence. I prefer to listen to them eating noisily, because that food is honestly earned. With them, everything is real. We need only observe, and feel.

That's why my team and I do not retreat into our own little miseries.. In our field of vision, the life of each passerby, even a faint ray of sunlight or a few heavy breaths can move us. We pay attention to the world around us, we feel other people's suffering. We care for what concerns them. We're not like those who avoid life's sorrows to bask in the radiance of reason. Nor do we immerse ourselves in the din of rock and roll, fixing our gaze on our own shadows and caressing ourselves. We genuinely want to understand others so that in this time when compassion and beliefs are losing ground we might engage with people's ways of thinking. We base our work on respect for the individual life, and we expose that life. We first concern ourselves with people's individual situations, then with the state of our society. We want art to convey truth and we want to uphold idealism. We're faithful to facts and to ourselves. We've made a commitment and I won't recant.

It is with this attitude that we point our camera at the city. And it is this attitude that makes us free, self-confident, and honest. In my opinion, adopting an attitude is much more

XIAO SHAN GOES HOME (XIAO SHAN HUI JIA), 1996

important than adopting a style. Our style can never be separated from the attitude we have toward the world. My attitude dictates my way of telling things, which then sets the entire tone and look of the film. Whether it was for *One Day, in Beijing* (*You yitian, zai Beijing*), *Xiao Shan Goes Home*, or *Dudu*,[1] I started with this basic tenet. It's the condition required for conversation and the way conversation works.

In *Xiao Shan Goes Home*, cutaway shots represent the characteristics of media. They reveal Wang Xiao Shan's place in the world, as well as our own. With the frenetic development of communications, we're surrounded and eventually worn down by media. Perhaps that's why we've become so cold. We're progressively losing our capacity for independent thinking and our ability to undertake face-to-face exchanges. Our ways of communicating have already been transformed. People have become used to exchanging with machines. Depressed, they listen intently to *Heart-to-Heart Talk*; they discuss social problems while *A Critical Moment* is on; they return to consuming after viewing *Worth Every Penny*.[2] People's lives are increasingly ruled by the market's norms and values. How many of these norms and values have not been fashioned by the media—the ever-increasing channels of China Central Television, the ever-expanding pages of *Beijing Youth Daily*? All of the above transform people. Wang Xiao Shan, who comes from Anyang, Henan Province, now lives among urbanites who have been remolded or who are being remolded by the media. The media reaches every corner

---

1. *One Day, in Beijing* (1994) and *Dudu* (1996) are short films that Jia made prior to *Xiao Shan Goes Home*. *One Day* (15 min.) shows people milling about on Tiananmen Square; *Dudu* (50 min.) showcases a lonely female student who is played by Jia's sound engineer, Lin Xiaoling, who was also studying then at the Film Academy.

2. *Heart-to-Heart Talk* (*Wuye qinghua*), *A Critical Moment* (*Jiaoji shike*), and *Worth Every Penny* (*Ming bu xu chuan*) were popular radio and television shows of the 1990s.

of the city, and Wang Xiao Shan is steeped in it.

Audio segments containing no visual component, where the story is told in a kind of radio broadcast, invite people to listen differently, without the visual support we have become used to. Segments that abandon both sounds and images, replacing them with text on a computer-like screen, highlight our new widespread practice of reading off of screens.

From written news reports and omnipresent advertisements to the editing style of CCTV's news program *Oriental Horizon*, the film's exhaustive combination of extant forms of media compels viewers to continually change and reflect on their way of processing information on and off screen, as well as the role of media.

Today, the pure exploration of film language no longer satisfies. But the search for innovative film forms is my true interest. That's why I worked hard on the graphic and architectonic levels of *Xiao Shan Goes Home*. I interrupted the linear unfolding of the plot with a magazine-style collage. In the assemblage of shots, sequences, and blocks, I felt the pleasurable rush that editors enjoy. It made me reflect on the true nature of cinema. While Hollywood uses seamless editing to deceive viewers, I, on the contrary, want to make editing all the more obvious and subjective. Editing that does not smooth over cuts or hide the frayed ends is more consistent with my goal of authenticity.

I still remember that day in the winter of 1994 when we rejected a whole slew of brash artsy names, like "Big Production" or "The Progressives," and chose instead to call our little group, which we had created on the spot, the "Youth Experimental Film Group." We still proudly stick to that plain name, which contains three words we like: youth, experiment, and film.

Originally published in *Avant-garde Today* (*Jinri xianfeng*), volume 5, 1997.

XIAO WU (A.K.A THE PICKPOCKET), 1998

2

Translated by C. Huot

SYNOPSIS

1997, Fenyang, Shanxi.
Xiao Wu is a pickpocket who thinks of himself as an artisan. He wears glasses with dark, heavy frames. He talks and smiles very little. His head is permanently cocked, his tongue tucked in his cheek. We watch him, in his Western clothing two sizes too big, trailing his hand along stone walls, rehearsing karaoke in the public bath, accompanying a female singer he's courting in vain on dull, boring walks through the city, or chatting idly with his former coworker who is now a big shot. In a small town undergoing massive construction, Xiao Wu wanders aimlessly.

NOTES

The camera captures the physical but targets the mind.

Behind the characters' empty chatter, vapid singing, and mechanical dancing, we discover the transience of real passion and the inconstancy of conscience.

This film is about the anxiety of living, how swiftly beautiful things disappear from our lives. In the face of failure and personal hardship, life once again becomes a solitary affair, tinged with a kind of nobility.

The film successively tackles the themes of friendship, romantic love, and family. What emerges is not so much a loss of feelings as a loss of codes. In the midst of chaotic streets, strident noise, and impermanent relationships, the characters seek escape by any means. And yet, listening to the toneless singing in the soon-to-be-demolished old buildings of Fenyang, we are somehow confident that, on the audiovisual plane, something is bound to happen.

ABOUT SOME SCENES FROM *XIAO WU*

1. By the highway, a youth and her family wait for the bus.

We were on the road at the end of a day's work when we chanced upon this scene by the highway: In the open countryside at the beginning of spring, a family is seeing off a young girl who is going far away. They are all silent, facing the equally silent tall mountains, staring down the highway.

I'm moved by farewells, so I shot the scene and put it at the beginning of the film.

2. In the pharmacy, Xiao Wu plays with Gengsheng's daughter. A white-haired policeman enters with demolition and removal surveyors.

Outside the window is the noisy county town. Inside the pharmacy, Gengsheng is giving Xiao Wu an earful while pulling his daughter away. That's when the white-haired policeman arrives. The policeman and thief chat, the conversation veering to Jin Xiaoyong's wedding.

Shooting this scene, I abandoned stopping the camera and simply let it pan back and forth. I suddenly realized that the pace of a county town cannot be easily segmented. So I watched, transfixed, from the sidelines.

3. Xiao Wu goes to Xiao Yong's home, the two sworn brothers have a face to face.

Xiao Yong is preparing for his wedding reception when he receives an unexpected guest, Xiao Wu. The latter plays with a lighter while the other fretfully looks on. This is a tense moment. After much deliberation, I decided against a dramatic flare-up. Who doesn't bury pain deep in their own heart? So Xiao Wu leaves the red envelope behind and exits alone, bringing his unexpressed feelings with him. These are the sorts of interpersonal relationships I recognize.

4. Xiao Wu and Meimei walk on the street of karaoke bars.

Meimei: "I shouldn't have worn high heels today."

Xiao Wu steps up onto the sidewalk.

Meimei: "Why don't you climb up to the second level—that's taller, no?"

Xiao Wu gracefully walks up to the skywalk level.

A kind of dignity, combined with spontaneity and introversion—these are the poignant traits of my friends from our county town.

5. On the second floor, Xiao Wu bites into a sour apple. Down below, his pubescent disciple and girlfriend are walking, one behind the other with bent heads in the noisy afternoon downtown. Music can be heard: a piece from The Killer.[1]

---

1. *The Killer* (*Diexue shuangxiong*) is a 1989 action film by Hong Kong director John Woo and starring Chow Yun-fat.

This is a moment I've often enjoyed during my twenty years of small-town life. On a bright sunny day, just after noon, gazing at familiar people and things. Then suddenly something swells up in my chest, everything feels brand new.

6. *Xiao Wu returns to see Mei Mei at the karaoke bar. The proprietress, carrying a basin of water, comes in from the courtyard.*

It turns out that this particular karaoke bar has an exit leading to the backyard. The proprietress comes in from the depth of field carrying water. This mundane sight clashes with the dimly lit karaoke bar, from which it is separated by a curtain. When I discovered this exit on location leading to the backyard, it gave me a thrill that lasted quite a while.

7. *Xiao Wu goes to visit Mei Mei who is sick. Throughout the scene, we see only partial views of his body.*

In spite of everything, he cannot fully enter this woman's life.

8. *Xiao Wu buys a hot-water bottle. He sits with Mei Mei, shoulder to shoulder on the bed. She sings Faye Wong's "Sky" for him.*[2]

The scene is backlit. The two characters, destined to part, are sitting together. Against the bright sunlight, their short-lived love appears nebulous. I often shoot against sunlight to give the oppressive world a bit of warmth—a world where love lasts but a fleeting moment.

---

2. "Sky" (1994) is an achingly romantic song performed by Chinese Hong Kong diva Faye Wong (Wang Fei).

9. Xiao Wu, kicked out of the house by his father, walks alone on a winding country road.

As the camera does a 360-degree panorama, the radio broadcasts a commercial of a villager selling pork, followed by news of the repatriation of Hong Kong. Everything close and everything far away gradually closes in on him; there is no option but to leave.

10. Closing scene: a crowd of onlookers surrounds Xiao Wu.

This ending was serendipitous. We were watching the film; the people in the film were watching us.

Originally published in *Today's Avant-Garde (Jinri xianfeng)*, volume 12, 2002.

PLATFORM (ZHANTAI), 2000

3

Translated by Claire Huot

SYNOPSIS

It's 1979. China has begun implementing its "Reform and Opening Up" economic policies.

Cui Mingliang, Zhang Jun, and several other young people from the Fenyang County Cultural Troupe, who perform song and dance, are on stage rehearsing their recitation of the poem "Undaunted" (Fengliu),[1] accompanied by instrumental music. The female performer reciting the poem, Yin Ruijuan, is Cui Mingliang's adored sweetheart. The two work together and often practice together. But their relationship is subtle; they've never spoken of it to one another. On a Sunday, Cui Mingliang and Zhang Jun make a date with Yin Ruijuan, Zhong Ping, and other coworkers to see the film *Awara*.[2] By chance, they bump into Yin Ruijuan's father, who doesn't like the idea of his daughter hanging out with Cui Mingliang or the possibility that they may be involved in a relationship. He tells his daughter to leave the cinema. Disgruntled, they all split up.

Fast forward to the early 1980s.

---

1. Young men and women memorized and bravely recited this long poem, written by Ji Yu and nationally broadcast in the early 1980s.

2. *Awara* (a.k.a. *The Tramp*) is a 1951 Hindi film by Raj Kapoor. This romantic story of a singing vagabond caused a sensation from India to Russia, including China.

At the hair salon, they listen to Teresa Teng[3] singing "Wine and Coffee." Zhang Jun takes time off to visit his aunt in Guangzhou.

Cui Mingliang receives a postcard from Zhang Jun in Guangzhou. He scrutinizes the skyscrapers on the card and has trouble sleeping all night.

Zhang Jun returns from Guangzhou with a digital wristwatch, a boom box, and a Kapok guitar. The Cultural Troupe prepares a light variety show to adapt to the times, and they get ready to hit the road. Yin Ruijuan's father falls sick, so she cannot join Cui Mingliang and the troupe for the tour. The young lovers must part.

Early morning: A bus carrying Cui Mingliang, Zhang Jun, and the others drives off toward distant lands. They have begun their tour.

---

3. Teresa Teng (Deng Lijun) (1953–1995) was born in Taiwan and died of asthma in Thailand at the age of forty-two. She was the most popular Chinese singer in Asia from the late 1960s to the early 1990s. Her romantic ballads were introduced to the Mainland from the South in the early 1980s.

NOTES

The film's timeline, from 1979 to 1989, covers a period of great change and reform in China. Those ten years were also the most important period of my youth. In China, the fate of the country is tied to one's own happiness; the political and the personal are always interconnected. During those ten years, because of the disappearance of revolutionary ideals and the imminence of capitalism, consumerism was on the rise. We experienced a lot during that period.

Platform is the title of a rock song from the mid-1980s that was trendy for a while. It talks of hope. I chose this song for the title of the film as a tribute to people's naïve expectations. A platform is the starting point, as well as the endpoint, of a destination. We are always expecting and seeking something, always on the road to somewhere.

Platform's narrative is structured by the stories of the characters as they change within their changing environment. In the natural course of birth, old age, sickness, and death, life has boundless sorrow in store for us. Flowers always die, and humans find themselves in situations where they have no options. At any rate, the theme of this film is human beings. I want to explore and display the hidden drive that compels humans to improve their lives. The film relates a period when the Chinese people shared common experiences. It's also, for me, a period I always recall with fondness.

ABOUT SOME SCENES FROM *PLATFORM*

1. The members of the Cultural Troupe file onto the bus at the end of their performance. The troupe leader begins the roll call. Soon after, he quarrels with Cui Mingliang. The bus sets out on its journey, gradually entering the night.

At this moment, there's a shift from the group to the individual. Five minutes into the film, there was still no way to identify the main protagonist, just as, in a collective, individuals are not what's important. It's only once the troupe leader makes the roll call and discovers that Cui Mingliang is absent that an individual is thrust into the foreground. This takes place inside a closed bus, and when the bus starts up it's like the beginning of a journey. For the people in the film, the bus is heading toward the future; for us—the viewers—the bus is heading back into the past.

2. The first time Yin Ruijuan and Cui Mingliang meet, they are at the foot of the very high city wall. That solid wall stands behind them. At their feet lies snow that has not yet melted. The sky is just a narrow band.

The camera begins some distance away from the characters and then retreats even further. I need the open space and the distance. I don't want to see their faces clearly, because they're standing in the cold of 1979. Little by little, a fire grows, but I don't want to insist on it, because the warmth, like gossamer, glows in their hearts.

3. After Zhang Jun's departure for Guangzhou, Zhong Ping, alone at home, flips through a songbook. Ruijuan arrives. The two of them, backlit, smoke.

Filming against the light, pointing straight at the window, the camera closes in on the characters. The two women's melancholy passes swiftly, as does their leisure time. This type of scene is exactly how I remember such moments. I get sad about the ceaseless passing of time. I felt a dull pain while shooting it. I was hoping the camera would just keep rolling.

4. At Zhong Ping's house: wild disco dancing to the beat of "Genghis Khan."[4]

The characters are enjoying the new times in a tiny, run-down blacksmith's workshop. Their excitement clashes sharply with their confined surroundings. The confined space also highlights the tension between ideals and reality. For the first time in the film, the camera shakes slightly. It remains at a distance from the characters. I like this type of contradiction in a scene—frenzy and restraint coexisting.

5. To the music of "Bella Ciao,"[5] *the walls of the county town become more and more distant.*

This is the first time the cultural troupe leaves on tour. It is also the first time the city walls and moat are shown in their

---

4. "Genghis Khan" was the German entry for the Eurovision song contest in 1979. The kids are listening to the Cantopop version sung by George Lam.

5. "Bella Ciao" is an Italian partisan song of the resistance during World War II that gained popularity internationally.

entirety—the way they encircle the city. After we completed this exit shot, I suddenly realized I had found the whole film's composition: "Enter the city, exit the city—departure, return."

6. Zhong Ping and Zhang Jun come out of Cui Mingliang's cousin's house. Zhong Ping says she feels like yelling. Zhang Jun says that's fine, so she squats on her heels and shouts at the top of her lungs a few times. A response comes in the form of an echo from the mountain valley. The camera turns away from the characters and fixes on the timeless mountain range.

When that happened, I felt an ache in my own heart—a kind of despair and emptiness.

7. Cui Mingliang's cousin chases after a distant tractor to give Mingliang a five-yuan [RMB] bill for his younger sister. He then turns around and leaves.

I'm amazed at the cousin's stride. It's so steady and firm as he returns to his harsh hand-to-mouth world. The actor playing the cousin is really my cousin. Shooting the film has brought us much closer. It was the first time I became aware of his body language, and of his dignity and self-confidence.

8. Ruijuan is by herself in the office dancing to the music on the radio. Next, she rides a motorcycle peacefully through the gray county town.

I don't want to provide reasons or explain why a young girl who was dancing is suddenly wearing a tax-clerk uniform or why she is still single after so many years. This is my narrative philosophy. Isn't that the way we get to know people and understand the world? In bits and pieces and on a superficial level? This is especially important when it comes to our understanding of change, even we ourselves don't understand when or where or why we've changed. What we see is the result. The result is all we can know.

9. After Mingliang and Ruijuan get married, they go to his father's auto parts shop.

I like small shops off the highway. When you stand there, you feel the road's vibrations.

10. Mingliang is sleeping soundly on the sofa; Ruijuan is holding their baby and pacing in the room. The kettle makes a high-pitched sound like a train's whistle.

Those who have lost their youthful vigor like to take a nap.

Originally published in *Avant-garde Today (Jinri xianfeng)*, volume 12, 2002.

IN PUBLIC (GONGGONG CHANGSUO), 2001

4

Translated by Claire Huot

SYNOPSIS

In a train station on the outskirts of town, a lone man dressed in a military coat paces back and forth. Late in the night, the sound of a train whistle is followed by the arrival of a train. A woman carrying a heavy sack of flour disembarks. At the bus stop in the mining district, it's dusk, almost nighttime. An old man struggles to zip up his jacket, a woman who missed her work-unit bus listens to the bell tolling several times as she gasps for breath. On a rattling and jolting bus, a young man is tormented by his toothache. A man with a pockmarked face enters a bus revamped as a restaurant. An old gang member in a wheelchair watches the girls go by. People are still singing and dancing, but the enjoyment is getting old…

NOTES

Wielding a digital video camera in public spaces, one encounters familiar strangers. After staring for a long time at crowds, I finally came into contact with arresting gazes. I pushed aside the identity of these people, their motivations and exchanges, to preserve only the gestures, sounds, and dejection and despair drifting in the dust.

IN PUBLIC IN MY OWN WORDS — 5

Translated by Sebastian Veg and Claire Huot

Last year I was awarded a grant from the Jeonju International Film Festival's Digital Short Films by Three Filmmakers program. Initially, I wasn't sure what sort of film I wanted to make. And then I thought of Datong. For me, Datong is a city of legend. Anyone from Shanxi will tell you that Datong is a really wild and scary place. So I thought I'd go and take a look. There was a particularly tantalizing rumor at the time that Datong was slated to be relocated, because the coal mines there had already been picked clean and the miners were all out of work and because this was the period of the Great Western Development policy. People were saying that all the coal miners were going to be moved to Xinjiang to work in oil drilling. In the meantime, according to the rumor, everyone in Datong was living it up while they could. Even in very ordinary restaurants, you had to reserve a table half an hour in advance. Datong and my hometown are at two ends of a diagonal line that cuts across Shanxi Province. Datong, in the northeast, is geographically and culturally closer to the cities of Hohhot in Inner Mongolia and Zhangjiakou in Hebei Province. Shanxi Province's capital, Taiyuan, is very remote.[1] And to add yet another exotic touch, Datong is a city of former grandeur on the decline; it has historic cultural sites attesting to its ancient and legendary past.

---

1. Here Jia Zhangke is making a pun: Taiyuan is a homonym for the phrase "too remote" (*tai yuan*).

I went there, my mind filled with fantasies. When I arrived, I became extremely excited, because it was exactly as I had imagined it. The bits and pieces of hearsay turned out to be quite similar to what I discovered, but the rumors were already quite outdated. This is how I became attracted to the city. At first sight, I felt the place was extremely erotic. Looking back now, this may have been because the people populating this contained space were all hyper-excited, full of desires, and bursting with life.

I did not immediately focus on public space. Initially, I merely wanted to conduct interviews. Because I had a friend running an all-night sauna, I thought of interviewing the people who spent the night there. I quickly abandoned this idea, because I felt that I neither needed people telling me things nor did I need to talk to them. I abandoned all language. Observing attitudes was quite sufficient. This is also the reason why the film has no subtitles. You don't need to hear clearly what people are saying; their voices are part of the environment. What they say is not important; what's important is how they look.

Then, as I was slowly getting into the filming, I found the right tone. Michelangelo Antonioni said something I particularly like: he said that when you enter a space, you must first immerse yourself in it for ten minutes and listen to what the space is telling you; then you can begin a dialogue with it. This has probably always been one of the tenets of my creative work: only when I stand in the space of the real situation do I discover how to shoot a scene. This is also how and when I compose my storyboards, so space and place are really extremely helpful to me. In a specific space, you can discover something; you get a feel for it that you can rely on.

I filmed many locations: the train station, the bus station, a waiting room, a dance hall, a karaoke bar, a billiard hall, a roller rink, a teahouse… During the editing, because of the film length limitations,[2] I had no choice but to cut out many things. In these locations, I found a kind of rhythm, a kind of tempo: many of these spaces were related to travel, so I selected the scenes that were best suited to this theme.

The atmosphere of a space provides an important direction. Another important factor is the link between spaces. I found it interesting to observe how the spaces of the present are superimposed on the spaces of the past. For example, a defunct bus was revamped into a restaurant; in the waiting room of a bus station, you could play billiards in front of the ticket counter; behind a curtain was a dancing hall. The space was divided into parts with distinct functions. It was like the superposition of identical images in some contemporary art. Observing the superposition of spaces, I discerned a deep and complex social reality.

The film became increasingly abstract, all the more so by the elimination of narrative and dramatic elements. In the end, I was left with a series of situations and details. What I hope will ultimately attract the audience are the faces, the characters.

This was my first time filming with digital video (DV). In practice, DV doesn't work exactly as I had imagined. Originally, I thought it would allow me to film very lifelike scenes, but I found that its best attribute was that it could even portray abstractions. Just as people tend to meander alongside a river in a particular rhythm, the advantage of DV is that you are part of the action, but you also keep an objective distance as you watch closely its rhythm and pulse while you move along with it. You conduct a kind of empirical observation.

We had a very small crew: a cameraman, a sound engineer,

---

2. When Jia was awarded the subsidy in 2002, there was a thirty-minute length restriction.

the producer, myself. We had one car. In the morning, we would drive around, just shooting whatever we saw in a very relaxed way, without any preparation. We explored as if we were just strolling. I think that's the kind of freedom DV gave us. Cinema is an industry, and making a film is a job that requires meticulous planning. Independent production is grounded in the desire to reduce, as much as possible, the constraints and fetters imposed by the industry. These constraints are not only the pressure imposed by a producer or the control of film censors but also the limitations of the filmmaking method itself, which has its own norms. DV gave us the pleasure of freedom from the industry. For example, we filmed the bus stop scene after our local guide had taken us to a coal mine to shoot a workers' club. When we came out—at the very place you see in the film—there just happened to be some people waiting for the bus. The sun was beginning to set, and I suddenly had the feeling I'd found the perfect shot. I started filming the place—and kept filming, filming, filming all sorts of things. By the time I shot the old man, I was already completely satisfied. He was very dignified, and I continued to film him patiently. When my lens followed him onto the bus, suddenly a woman appeared. My sound engineer later told me I had started trembling at this point. Watching her with the workers' drab living quarters as a backdrop, I suddenly had an almost religious feeling, and I kept filming. Then another man came into the picture—I don't know what sort of relationship they had—and the two of them left. During this whole sequence, I felt that every minute was like a gift from God.

This was made possible by the way we worked. I was able to spend a whole afternoon there just filming. But if I had been using film, it would not really have been possible to catch all these things. If you need to think about the ratio of film you are shooting to what you'll be using, then once you've more or less captured the scene, you better move on. But now I can film entirely as I like, because the way I work is very inexpensive.

But DV also has technical limitations. I was using a Sony PD 150 camera. My main problem with this camera is the focus limitations. It has no gradations, so it goes out of focus easily and you need to pay constant attention. Also, when you're filming outdoor scenes, the colors are terrible in strong sunlight and the depth is not very good. If you encounter any sort of shiny surface, especially a metallic horizontal surface, because the wavelengths are different, it will flash and the picture becomes unstable.

I once wrote an article entitled "The Age of Amateur Cinema Is about to Return."[3] After it was published, there was a great deal of discussion and, I think, some misunderstanding. I meant to argue for films made in an amateur spirit, as opposed to stale filmmaking, especially within the system. But with regard to the films themselves, our standards should be extremely high. The images we can capture on DV definitely represent a revolutionary, subversive step forward, but gaining easier access to the right to film doesn't mean a filmmaker should treat the work lightly or ignore quality.

In 1995, I shot a documentary called *One Day, in Beijing* (*You yitian, zai Beijing*), which I never edited. But when I reverted to documentary after making two feature films, I discovered I had developed a method. When you pick up the camera and begin shooting, the work itself forces you to experience spaces, people, and events that you have no opportunity to experience otherwise. I believe a director gradually becomes less self-confident, and less self-confidence does not translate into bad filming. On

---

3. A translation of this seminal text is included in Part IV of this book.

the contrary, when you are depicting someone or something, you experience doubts: "Is this what the person is thinking?" "Are this person's values, questions, and ways of dealing with problems really the way I'm representing them?" If I stay long enough in a particular environment, I can completely put myself in someone else's place—Xiao Wu, or a song-and-dance troupe—but after being away from them for a while, I can no longer be sure because I've already lost touch with their lives.

Making a documentary can help a director come out of this mindset. When I feel that my life is changing more and more, that my thirst for knowledge is weakening, and the resources of my life are dwindling, shooting a documentary gives me new vitality, as though my blood were beginning to circulate again after my arteries had been clogged for a long time. Because making a documentary enables me to reposition myself, I become grounded again, re-energized by real, down-to-earth life.

First published in *Art World* (*Yishu shijie*), December 2001.

UNKNOWN PLEASURES (REN XIAO YAO), 2002

6

Translated by Claire Huot

SYNOPSIS

Binbin stands staring into space in the waiting room of the bus station. He's not going anywhere. He's never even considered leaving this city. Xiaoqi sits on the bench facing the ticket office, smoking and reading a newspaper. He's not reading the news; just looking for a job.

They're friends; they don't care for talk, but they enjoy hanging out together.

This city is called Datong and the popular song of the moment is "Free as the Wind" ("Ren xiao yao").[1] Datong is northwest of Beijing, far from the sea and close to Mongolia. "Free as the wind" is an old phrase that Qiaoqiao understands as meaning "just do what you wanna do."[2] Qiaoqiao's nickname is "White Thighs"; she's the city's famous freelance model.

This year they all talk less and less, but there's clearly some kind of excitement below the surface of their silence. This year, a deafening sound tore through the city sky. They couldn't tell if it was thunder rolling in from far away or waves breaking in their dreams.

The year is 2001. They are nineteen years old.

---

1. A 1997 song by Taiwanese pop singer Richie Jen used by Jia for his film's title in Mandarin. The English translation of the film title comes from yet another song, "Unknown Pleasures" by Joy Division (1979).

2. *Xiaoyao* is a term especially associated with the Daoist philosophy of Zhuangzi. It refers to the unobstructed course of nature.

NOTES

Standing on the streets of Datong, looking at the young people's apathetic faces… With the arrival of globalization, this gray industrial city is looking sexier by the day. People try hard to be happy, but the faint smell of gunpowder pervades the night, along with songs from karaoke bars.

Bankrupt state-run factories are everywhere in this city, but all that's produced here is despair. Young people walk around with closed fists. They're the children of unemployed workers and in their hearts there's no tomorrow.

With my camera I converse quietly with the city, and I slowly realize that the carnival atmosphere is rooted in utter despair. So I become like them, excited for no particular reason.

Can you see it? Violence is their last romantic stand.

THE WORLD (SHIJIE), 2004

7

Translated by Alice Shih

SYNOPSIS

Zhao Xiaotao is making a call on the monorail. She says she is heading for India. Her ex-boyfriend suddenly wants to meet up with her. He said he is going to Ulan Bator.

What Zhao Xiaotao refers to as India is actually a section of a theme park. She works there as a dancer for the tourists. Ulan Bator, where her ex-boyfriend is heading, is the capital of Mongolia, far north of Beijing.

They meet and eat. The small restaurant is filled with smoke, shrouding over their sadness of separation.

Zhao Xiaotao is talking on the phone on the monorail again. She says she wants to meet with him. "Him" is Cheng Taisheng, her current boyfriend. He is the head of security in the theme park, and he is working at the Eiffel Tower right now.

They all live inside the park; working, eating, wandering, and arguing with one another within this community.

They all come from other parts of China; they fantasize, fall in love, doubt, and reconcile here in this city.

This is Beijing in 2003. Noises drown out the whole city, exhilarating some people and silencing some others.

This theme park is filled with miniatures of tourist attractions from all over the world. It takes only ten seconds to travel from the pyramids to Manhattan. In this replica of sceneries, reality creeps up on them. One day could feel as long as a year. The world could be as small as a corner.

NOTES

1. The Village

After finishing *Unknown Pleasures*, I went back to Fenyang, my hometown in the province of Shanxi. I visited my cousin in a very small village. He continued to work in the only coal mine there. He appeared to be more lonesome than before, possibly because most men his age have left for a bigger city or to the south where the economy is in better shape. Left behind them are seniors, kids, and disabled people. This village is in the middle of nowhere, you hardly see any people on the street. My cousin asked about Beijing. He must be missing his friends who left to live there. His question brought me back to the hustle and noise of the busy streets in Beijing. I thought of the faces of the newcomers who appear before me daily. I thought I should make a film about Beijing for him, as it would be very difficult to explain everything through words alone.

2. Scenes of the County

A. Once it hits ten o'clock at night, Fenyang goes dark. To conserve energy, the streets are only lit up on the left side.

B. The Internet bar is busy. With a girl on his arm, a young man taps on the keyboard and drinks a beer. The bar is also selling noodles. The backyard has some guest rooms, equipped with stove-heated beds for people who take breaks from the virtual world.

C. In the cultural center, a young man is skating proudly by himself in the rink with a can of Coke in his hand. He drinks effortlessly from the can while he skates. The Coke can is an important prop for his image. It is what he uses to differentiate himself from the crowd.

3. Beijing

I had planned to meet a friend at Jiangnan Restaurant located at the northeast corner of the Dongsishitiao overpass. When I got there, I found my friend waiting for me in front of a pile of rubble. A few days ago this was a high traffic area, but now Jiangnan Restaurant and its neighboring stores were all gone, reduced to ruin. I came to realize that surrealism is Beijing's new reality.

This city has been transformed into a vast construction site over the past few years. It has also turned into a giant supermarket and an enormous parking lot. On one hand, the public is provided with a variety of entertainment venues. On the other hand, tens of thousands of people are losing their jobs. Skyscrapers continue to rise up while people are financially ruined. Migrant workers sacrifice their health and lives to light up the city's neon lights at night, and each morning the city wakes up to greet another crowd of newcomers. The difference between day and night in the city is getting smaller, and the seasons are becoming indistinguishable. The pace of life is fast, but we have lost leisure.

4. The World (the Theme Park)

In 1993, I toured Beijing with my parents. I remember traveling through heavy traffic from the city center to the remote outskirts to visit the theme park called the World.

There, the exotic architecture we had only ever seen on wall calendars was right in front of our eyes. Passing through the Egyptian pyramids and the U.S. White House, we saw the Red Square of Moscow. Cacophonous bells on Indian lady dancers were juxtaposed with decorous Japanese music. "Travel the whole world without leaving Beijing!" was written on a billboard. It was all a convenient and simple way to satisfy the curious heart's passion to learn about the outside world.

Interest in the World's replicas reflects people's passion to know the unknown world, but the whole experience is a misperception of reality. When they stand in front of these meticulously reconstructed replicas, people should realize that they are even further away from the real structures they represent. I thought of those who lived inside the theme park. While it seemed they could roam freely without borders, they were, in fact, being closed in on all fronts. People may be able to replicate the Eiffel Tower, Manhattan, Mount Fuji, or the pyramids, but the lifestyles, the social systems, and the cultural traditions of these places remain inaccessible. You still have to face your own pain in your own world. Enjoying the fruits of globalization cannot rewrite the past. Postmodern scenes cannot conceal existing problems that are the residue of an earlier age.

The World

I want to make a film about *The World*.

The more I think of it, the more I believe that "one day can feel as long as a year; the world can be as small as a corner."

"Ulan Bator by Night" (featured song in *The World*)
Lyrics by Jia Zhangke and Zuoxiao Zuzhou

The wind that cuts across the plain
Slows down.
My silence is telling you
I'm drunk.
The cloud that drifts in the distance
Slows down.
I'm fleeing to show you
I'm not turning back.

The nights in Ulan Bator,
So quiet, so quiet,
Not even the wind knows that I don't know.
The night of Ulan Bator,
So quiet, so quiet,
Not even the cloud knows that I don't know.

Travelers who roam far from home
Where are you?
My stomach starts to hurt.
Do you even know that
Birds that fly through fire
Do not want to go?
You know those who go mad on this night
Are not alone.

The nights in Ulan Bator,
So quiet, so quiet,
Not even the wind knows that I don't know.
The night of Ulan Bator,
So quiet, so quiet,
Not even the cloud knows that I don't know.

STILL LIFE
(SANXIA HAOREN), 2006

8

Translated by Alice Shih

SYNOPSIS

Coal miner Han Sanming travels from Fenyang to Fengjie to look for his estranged ex-wife of sixteen years. The two people meet on the banks of the Yangtze River, lock eyes, and decide to re-marry.

Nurse Shen Hong comes from Taiyuan to Fengjie to look for her husband who hasn't come home in two years. They hold each other tight in front of the Three Gorges Dam. They dance sadly and agree to divorce.

The old county is already submerged in water, and the new site is still under construction. Things to be kept should be picked up, and things to be discarded should be left behind.

NOTES

I entered an empty room one day and saw a bunch of dusty objects on the table. I seemed to have uncovered the secret of still objects. Those furnishings that sat in the same places for years, the dusty utensils on the table, the wine bottles on the windowsill, and the decorations hanging on the wall all carried some sort of poetic sadness. Still objects are one part of reality we neglect. Although they endure, they remain silent and hold secrets.

This film was shot in the old town of Fengjie. This whole area has been hit hard by a series of big changes as a result of the construction of the Three Gorges Dam. Families who have lived here for generations have been forced to move away. This old town with a history of two thousand years will be demolished and submerged under water in two years.

We brought our camera into this soon-to-vanish town. We see demolitions, explosions, and rubble everywhere. Among the deafening noise and flying dust, there was still a sense of vibrant color blooming out of life itself, despite the despair.

Teams of workers running back and forth in front of the camera. I'm in awe of the stillness of their silent expressions.

DONG, 2006

9

Translated by Alice Shih

SYNOPSIS

Fengjie, China, 2005
Painter Liu Xiaodong went to the Three Gorges region to paint *Warm Bed* (*Wenchuang*). Twelve demolition workers became his models. This two-thousand-year-old town will soon vanish with the construction of the Three Gorges Dam. Spending time with his models, the painter is struck by the reality around him.

Bangkok, Thailand, 2006
He paints the second part of *Warm Bed* in Bangkok, using twelve tropical ladies as his models this time. The intense heat of the city makes the ladies sleepy, but the fruit on the floor still looks colorful. The painter overexerts himself and gets tired. The ladies open their eyes and sing a cheerful song.

    Rivers run through both places. The water streams forward without looking back.

NOTES

Bringing a camera, we followed Liu Xiaodong into the town of Fengjie, which was in the process of being demolished. We also followed him to the city of Bangkok in the heat of summer. We got to know twelve men and twelve women who lived in these two places, thousands of miles apart yet facing similar circumstances. This casual journey with the camera unveiled so much of what Asia is about.

USELESS (WUYONG), 2007

Translated by Alice Shih

SYNOPSIS

Summer in Guangzhou:
A fan is blowing on dresses hanging from wires in muggy Guangzhou. We see the faces of garment-worker women through the gaps. Surrounded by the roaring noise of sewing machines, workers labor quietly under fluorescent lights. No one knows who will wear the newly made clothes, and no one knows what the future holds for these faces at the conveyor belt.

Winter in Paris:
Guangzhou fashion designer Ma Ke is launching her new brand, Useless, at the 2007 Paris Winter Fashion Week. She buries her stylish clothes in the dirt, letting nature work with time to finalize the design. She likes to convey emotions through handmade clothes. She's tired of conveyor belt productions and shuns machine-made fashion.

Dusty Fenyang, Shanxi:
A small, remote tailor shop in the mining district is occasionally visited by miners. They come for repairs and to chat. Flickering mining lanterns and cigarettes glowing between their fingers exude the same loneliness. Newly darned clothes in their plastic bags bear a trace of warmth.

NOTES

Tracing the path taken by the clothes, we shot at three different locations and discovered the dissimilar realities that exist along the same chain of production and consumption. Clothes cover a body, convey emotions, and carry messages. Clothes stick to our skin and have their own memories.

24 CITY (ERSHI SI CHENGJI), 2008

11

Translated by Alice Shih

SYNOPSIS

It is 1958, a factory in the coastal Northeast is being relocated inland to the Southwest. Dali (played by Lü Liping) is a first generation worker who traveled that same year from Shenyang to Chengdu. The long migratory journey has been hard. Xiaohua (played by Joan Chen), who goes by the nickname of "Standard Piece," was dispatched from the Shanghai Aviation Academy to this factory in 1978. She is considered the beauty of the factory, which is how she got her nickname that implies she "sets the standard of beauty." Nana (played by Zhao Tao) was born in 1982. She walks between the modern city and the old factory and claims to be the daughter of a worker.

The stories of these three generations of women linked to the factory combine with the experiences of five other narrators to show what happens at a state-owned factory in the process of shutting down. Their destinies unfold there.

In 2008, this factory will move to another industrial area. The site, which is right in the heart of the city, is being purchased and will be developed as condominiums named 24 City. Things in the past become memories as stars travel onward through time. This is something strange yet familiar. Efforts to build up from the past are respectable, but the need for urbanization of the city today is also justified.

NOTES

This film is made up of the fictional narratives of three women and the retelling of five real-life experiences. The best way I could think of to show the history of China from 1958 to 2008 was to use documentary and mockumentary forms together. From my point of view, history is built with facts and fabulation.

The story takes place in a sixty-year-old, state-owned military factory. I was not too concerned with the historical details of the factory itself. I just wanted to listen and understand the personal experiences going on in a place going through so much historical and social change.

Contemporary films rely too much on action. In this film I wanted monologues and narrations to be the only forms of action—to let words alone reveal the complicated experiences buried in the subjects' hearts.

It's not important whether the factory was a happy place or a difficult place; we can't neglect the people who spend so much of their lives there. There are eight main characters in the film, and I hope that everyone will be able to relate to at least one of them.

PART II
THE MAKING OF A "GRASSROOTS" DIRECTOR

I DON'T ROMANTICIZE MY EXPERIENCES   **12**

Translated by Claire Huot

I was in the second floor café of the Joint Publishing bookstore waiting for someone when all of a sudden a bevy of artists came in. They were all around forty years old, longhaired, and wearing the "uniform." They gestured broadly and advanced as if they were the only ones on earth, loads of panache.

The leading elder of the group sat down and began lecturing on cinema. He spoke like a priest, evangelizing as if each and every sentence were truth itself. When mentioning a person, he would drop the family name, often saying "Kaige" for Chen Kaige, "Old Mou" for Zhang Yimou. Everyone around the table listened intently in awe. He said: "That gang of youngsters don't get it; they haven't suffered; they haven't experienced anything; how can they make good films?" He followed this up by reminding them how "Kaige was a sent-down youth" and "Old Mou had to sell his blood." As though only those experiences qualified as real experience and this kind of suffering was the only true suffering.

Our culture encourages that sort of worship of "suffering," and it has obtained capital in the dominant discourse. Consequently, by force of habit, some people make a habit of proclaiming their suffering, considering that only their own experience qualifies as suffering. But what does other people's, the next generation's, suffering count for? They have had only few bumps along the road. The rest of us must never compare our suffering and experiences to theirs. We must shut up. Suffering is a privilege of the few, and it has become the main way to evaluate people.

This reminds me of the public sessions when people shouted, "Recall the bitter past in the light of the sweetness of today!"[1] In those days it was assumed that suffering was a thing of the past and happy times were here and now. Who would have thought that during those "happy times" we were experiencing a calamity? The younger generation is not necessarily happier than the older one. Everyone knows that that thing called happiness does not, like material goods, increase with each passing day. I don't believe that a child dumped in a room and propped up in front of the television by his parents is necessarily happier than the previous generation's youth covered in sweat while they harvested wheat under the sun. Everyone has his or her own problems; each generation has its troubles; no one is worse or better off. Suffering is equally distributed.

In the contemporary poet Xi Chuan's words: "A crow solves a crow's problems; I solve my own problems." With this independent spirit in mind, we can watch *Beijing Bastards* (*Beijing zazhong*) and understand Zhang Yuan's anger and dysphoria. We can also understand the feeling of alienation in Wang Xiaoshuai's *The Days* (*Dongchun de rizi*). But the cheerless atmosphere in Zhang Ming's *Rainclouds over Wushan* (*Wushan yunyu*), or He Jianjun's gloom in *Postman* (*Youchai*) are manifestations of each director's personal raw pain. These filmmakers are no longer attempting to speak on behalf of a generation. Indeed, no one has the power to represent others; you can only represent yourself. That's the first step in shaking off culture's shackles. It's a

---

1. Those public sessions, *yi ku si tian*, were held during the Cultural Revolution. People would report on their pre-1949 hardships (*ku*) in contrast to their good lives (*tian*) in New China.

sort of wisdom and, even more, a way of life. Hence, for them, "suffering" is strictly a personal affair. If you don't grasp that, you can't enter their emotional world. Often, I've noticed that people go to see movies that suit their expectations. If the film differs from their experience, they're alarmed and then scornful. It isn't in our power to judge another person's life. I really like the title of one of Werner Herzog's films, *Even Dwarfs Started Small*. While life is not filled with that many extraordinary events, each of us experiences so much while growing up.

Unlike the Monkey King, no one was born and immediately leapt off of a stone crevice. I've come to doubt the older filmmakers' understanding of experience and suffering. In our culture, there are always people who romanticize their own experiences, who invent fantastic tales about themselves. As if ordinary mundane life was not enough for these superior beings, as if one must endure terrible calamities and misfortunes and encounter extraordinary hardships to be able to understand human affairs. To create such personal legends is deliberate mystification.

What I want to point out here is that this mindset has seriously harmed Chinese cinema. Some people make films that seek the extraordinary: they weave together tragic and happy events, great celebrations and great tragedies, as if that were the only stuff of film. When tackling society's actual complexities, they get flustered and confusedly churn out endless puerile nursery tales.

If cinema is going to show concern for ordinary people, one must first have respect for everyday life. One must follow the slow rhythm of life and empathize with the light and heavy things of an ordinary life. "Life is like a long, calm river";[2] let's experience it.

---

2. Jia is paraphrasing the Chinese title of the French film, *La vie est un long fleuve tranquille* (1988) by Étienne Chatiliez.

In an essay, Bei Dao wrote: "People always believe that the windstorm they've endured is unique; and that they themselves have become a windstorm to shake up the next generation."[3] He concludes: "How will the next generation live? It's up to them to answer." I have no idea of how we'll live or the kinds of films we'll make. "We" is a meaningless word. Who is "we"?

---

3. It can be found in his collection of essays *Blue House* (*Lan fang zi*), trans. Ted Huters and Feng-ying Ming (Brookline, MA: Zephyr Press, 2000).

I HAVE MORE HEADACHES THAN THE MONKEY KING[1]

13

Translated by Claire Huot

In my third year of junior high, one of my sworn brothers was leaving to become a soldier. I went to say goodbye. He was sitting on the couch and didn't say a word. I wanted to comfort him so I said, "You'll be fine; you're free now!" Even after I said this, he still looked miserable and even a little annoyed. In my mind, if you didn't have to go to school, do homework, or be pestered by teachers, you couldn't be happier. And to be able to leave home for faraway places, free from your parents' discipline? That made it an all the more happy thing. I sincerely envied him. I couldn't understand why my buddy was so sullen, his head bent down, not even looking my way. As the coach was leaving, I didn't want to see that weepy face anymore, so I said, "Are you faking being unhappy just to make us feel better because we have to watch you fly away and we're upset?"

My friend raised his head and placidly said, "'Free,' my ass! I'm off to military service!"

I said, "Looking at that weepy face of yours, you'd think you were going to do forced labor."

He took his time before he replied, "Balls! The army, forced labor: it's all servitude!"

I was dumbstruck. Here was a guy who didn't even like to

---

1. The legendary Monkey King, a.k.a. Sun Wukong, wears a magical headband that cannot be removed. If he goes astray, the band tightens and causes him unbearable headaches.

listen to storytellers; all he cared for was his couple of pet rabbits. But now that he was enlisting, the words he'd just spoken seemed to come from another person. His words flashed like lightning, lighting up my world. I was still in the midst of puberty; my voice had barely changed. I'd never questioned anything; I took everything for granted. But suddenly my friend seemed to have developed a capacity for analysis. With a single sentence he'd not only surprised me, he'd lifted me right out of my own age. I'm still grateful to him. For that instant he'd become a sage. Inadvertently, he opened my mind in a way that would benefit me for the rest of my life. To this very day he has no inkling of what he did.

It was a huge life changer for a kid who previously only thought of his rabbits. He was now reflecting on the true face of freedom. His view on the question of the army transcended his own physical environment, the value judgments of his society, and the unquestioning acceptance of those values in the culture of 1984. He had seized on the root of the problem. His statement, so broad and full of humanity, shocked and freed me.

My friend's insight cracked open my own reflexive thinking. On my way back home after seeing him off, I pondered the question of freedom, whether it actually exists in the world. You might escape from your family but then there was school. If you were free from school, there would probably be something else. Outside the law were more laws. All of a sudden, I seemed to have understood the significance of the Ming Dynasty novel *Journey to the West* (*Xiyouji*). With one somersault, the Monkey King could travel a distance of 108,000 *li* (34,000 miles). Free as the wind, he knew no spatiotemporal restraints, but he could not free himself from the palm of Sakyamuni's hand. This gave the Monkey King headaches, and no incantations could dispel the law. The author, Wu Cheng'en,[2] was a tragic figure of the Ming Dynasty, and his *Journey to the West* was a veiled criticism. There is no resistance, only struggle throughout the novel; no freedom, only laws. Of all the classical Chinese novels, *Journey to the West* is the one I can most relate to. Of course, this is only my personal reading.

The problem of freedom is one source of humanity's pessimism. Pessimism creates a climate for art. Pessimism brings us down to earth and makes us decent. It fills us with creativity. One reason art exists must be to expose our lack of freedom. That feeling is not limited to a particular period or generation, and it is certainly not restricted to any particular ideology. The lack of freedom is a basic human experience, like birth, old age, disease, and death. To a certain degree, art reflects our understanding of freedom: since we can't escape laws, does this imply we must abandon the quest for at least limited freedom? In other words, we can attain relative freedom by adopting the down-to-earth position that springs from pessimism. This is the way to get closer to freedom.

After graduating from high school in 1990, I didn't take the entrance exams for university. I didn't feel like studying. I wanted to go work and earn some money. I thought that if I became financially autonomous, with no need to rely on my family, I'd have some freedom. My father was vehemently opposed to this plan. He had not been able to go to university, because of his "bad" family background, and he wanted me to fulfill his dream. I felt my family was denying me my freedom. I had no desire to be somebody special. What was wrong with playing mahjong,

---

2. Wu Cheng'en (ca. 1500–1582 or 1505–1580) authored the novel *Xiyouji* (*Journey to the West*), which is considered one of China's four great classical novels, along with *Water Margin* (*Shuihuzhuan*), *Romance of the Three Kingdoms* (*Sanguo yanyi*), and *Dream of the Red Chamber* (*Hong lou meng*).

hanging out with friends, and watching TV? I didn't feel that anyone's life was nobler than any other's. My father said that I had a bad attitude, that I was negative and passive. But what was wrong with that? To my mind, Marcel Duchamp and the Seven Sages of the Bamboo Grove³ were all slackers. Duchamp played chess all day long; why couldn't I play mahjong? I wasn't harming anyone. Several years later, when I made *Xiao Wu*, the authorities deemed that the work was full of negativity, which reminded me of my father. They all thought like children's guardians.

And then I went to Taiyuan to study art in a pre-university class administered by Shanxi University. That was the result of a compromise between me and my father. To get into an art school you didn't need to pass a math exam, which was a way to avoid my weakest subject. I was able to leave home, go to the city of Taiyuan, and enjoy my freedom.

At that time I had a classmate who was working in Taiyuan. He'd begun to earn money and was financially autonomous. I envied him, and one day I went to see him at work. To my surprise, none of the office workers had left, even though the shift was over. They were there with their section chief, playing poker. If the boss didn't leave, no one left. It wasn't easy to get away, my classmate said, and he was bored to death of poker. I suddenly saw the scary side of power. I said to him, let them play their game, you tell them you have business to attend to, won't that do? He said he couldn't do that; if he didn't stay to play poker with the chief, he wouldn't be part of the team, and how would he survive? I realized my friend was already playing the game. He had his reasons, but from that moment I lost interest in a regular job.

---

3. The Seven Sages of the Bamboo Grove (*Zhulin qi xian*) were non-Confucian eccentric poets and musicians of the third century who enjoyed alcohol and celebrated nature.

Not to even realize the loss of one's freedom can be called ignorance; but to be aware of that loss of freedom and not to attempt to get it back can be called a lack of courage. Human intelligence in this world is surely not lacking, so what's lacking has to be courage. Choosing a way of life involves freedom. Rebelling against a way of life also involves freedom. Beginnings and endings involve freedom. Freedom requires us to make decisions without consideration of gain or loss, without fear of pain.

After I read *Journey to the West* as a youth, I stood by myself in the yard looking up at the open blue sky, muttering incantations, hoping that those words were the right ones to let me fly freely across space with a single somersault that would propel me 108,000 *li*. I fell a lot in those days, and I never succeeded in flying. I often think that my headaches are worse than the Monkey King's. He can fly, reach the sky, and come back to the world of the living, but I definitely can't. I have to bear my destiny. Nothing is new in a sunset; from the sun's perspective it's old hat, but for me it's new. So it's best I keep on making films. This is my way of approaching freedom.

I'm pessimistic, but I'm not alone. In terms of freedom, even the Monkey King is in the same boat as we are.

Originally published in *Satellite TV Weekly* (*Weishi zhoukan*), 2003.

WHEN ONLY WINE CAN UNLEASH YOUR STREAM OF CONSCIOUSNESS   14

Translated by Claire Huot

After *Xiao Wu* came out, I was suddenly getting a lot more invitations to go out. I didn't want to slight or miss anyone. When you're out to make it in the wild world of showbiz, the more friends you have, the more inroads. Someone like me, who arrived in Beijing with one suitcase, searching for a way in, is immensely grateful to be noticed. Knowing people is far more important than knowing places. Also, in those days, I had loads of time. I was happy to be in the company of others, even if only to make small talk.

Where to meet was a problem. I didn't have an office, and my place was too tiny and messy to receive guests. Whenever I had a meeting, I'd let the other party decide the place. The guests, equally polite, wanted to please me. So they'd fall silent for a moment, thinking, until they came up with the same three words: Huang Ting Zi (the Yellow Pavilion).

Even today, my zone of activity remains within the area we call Xin Ma Tai: "Xin" for Xinjiekou, "Ma" for Madian, and "Tai" for Taipingzhuang—all places close to the Beijing Film Academy. After four years of studies, I know them very well: my legs carry me there on their own. Huang Ting Zi, short for Huang Ting Zi Number 50, is a bar situated a few hundred feet north of the Beijing Film Academy. Because the Children's Film Studio is across the street, it's easy to find. In the afternoon there are few customers, so it's a convenient spot to have a conversation.

Back then the outdoor restaurant near Beihang[1] was in its heyday. At nighttime, people from all walks of life swarmed the place. Through the thick smoke and scent of cumin emanating from the lamb stew, you could witness fights with sticks and bats. In the back, "Uncle" from Xinjiang would call the waitress from Sichuan in a mix of Mandarin and Uighur. Turning around, you'd see a table of university students inexplicably crying. This place was wild and disorderly, but it was full of vitality, which suited me.

But Huang Ting Zi was a different outfit altogether. Every time I passed by on my way back to school after having been to Beihang, my heart would sink when I saw the dim lighting through the windows. Shanxi people are poor: from as far back as I can remember, my parents saved electricity. I've had my fill of gloomy fifteen-watt lighting; I love brightness. As a young man, with my head full of stories to tell, the romance of candlelight was never my thing.

I first stepped into Huang Ting Zi in 1997. I'd met cinematographer Yu Lik-wai in Hong Kong and we agreed to make a film together. He arrived in Beijing before I'd written the script. He phoned from Huang Ting Zi, where he was waiting for me. When I got there, there were already a few empty bottles on his table. I lit a Dubao cigarette, which is often called "Derby," which connotes bad luck.[2] But our meeting went so well that we decided, based on that single meeting, to go to Shanxi together, and that's how *Xiao Wu* was born.

---

1. *Beihang* is a contraction of Beijing hangkong hangtian daxue (Beijing University of Aeronautics and Astronautics). Like the other places mentioned, it's located in Haidian District.

2. The Chinese name for Derby cigarettes is Dubao, which approximates the sound of the English and means "Capital Pleasure" in Chinese. When the English name "Derby" is pronounced, people hear a slang word for "bad luck," "*dianrbei*."

WHEN ONLY WINE CAN UNLEASH YOUR STREAM OF CONSCIOUSNESS

Little Yu is a good drinker and became a fixture at Huang Ting Zi. That's how I ended up there frequently and made friends with the owner. After a while, people joked that Huang Ting Zi was my office. The owner, Jian Ding, is a poet, and his bar was also a poetry club. Around midnight he'd force Little Chen to play Chinese chess with him. Little Chen was the bartender, and when he saw me come in, he'd greet me as "Big Brother Jia" and tell Lili to serve me tea. Lili was the waitress, a distant relative of Jian Ding. She loved to watch television and wore her hair in small braids and made herself up like an actress from the Republican era. I had found a new place to haunt; even if I had no one to meet there, I always had someone to talk to at the Huang Ting Zi. There were many people like me, including a Brit called David who taught at the Beijing University of Chemical Technology. He would arrive promptly at midnight, order a draft beer, crane his neck to watch the soccer game, and chat with Little Chen about David's hometown, London. As these homesick faces came together by chance at midnight, there was no real friendship, and so there was room for real talk.

I still meet people at Huang Ting Zi in the afternoons. We raise a reminiscent cup, settle accounts with enemies, interview each other, talk about productions, plead for help, and get pointers from experts. I don't drink much, but I talk a lot. My hometown, Fenyang, produces Fen wine, which has been celebrated by some famous people. One night, I suddenly recalled a poem, written by I can't remember who. It says, "Now only wine can unleash your stream of consciousness / happy, though you've not finished your work."[3] That got me thinking. As I poured wine and exchanged toasts, my heart sank: I knew I was not doing what I

---

3. These are lines from a poem by Wang Meng (b. 1934), a contemporary writer famous for using stream-of-consciousness in his works.

ought to be doing. I began to feel bad. I stopped talking, leaned over the table, and stared at the flickering candles. The ambient sounds slowly turned into white noise. I was cloaked in the atmosphere of *Flowers of Shanghai* (*Hai shang hua*).[4] I thought about getting old and that I'd outgrown the time for muddling along. My life was frivolous and my body heavy. Like an old man, I got up abruptly. On my way home in the dark night, I revisited my childhood. I knew I was tipsy and said to the driver, "Now only wine can unleash your stream of consciousness." The driver, having seen many like me before, did not respond, because he knew that, in the light of day, his passenger would snap out of it. I'd joked and laughed along with anybody, shook hands with everybody, never realizing what a fawning buffoon I'd become.

The next afternoon, I was once again waiting for someone. My guest arrived late. I wasn't as restless as I'd been the previous night. I let myself go along with the afternoon's leisurely pace and went to the window to look outside. People were out in the sun, riding their bicycles, hustling and bustling about. Who knew what they were chasing. I thought, ordinary people are like sparrows, and it made me sad. Suddenly, a middle-aged woman came in. She asked Little Chen for a drink and if he could put on a song by Jeff Chang.[5] When the song started, she began to weep. This bar was also a place where people came to cry.

If you go back to Huang Ting Zi today, the bar has been demolished. All that remains is a heap of dirt. That's a metaphor for everything. Everything turns into dust. That's why I must devote myself to film. Not for posterity, but for all the tears that are shed.

Originally published in *Satellite TV Weekly* (*Weishi zhoukan*), 2003.

---

4. *Flowers of Shanghai* (1998), by Hou Hsiao-hsien, takes place in a brothel. The lighting is low-key and the mood is stifling.

5. Jeff Chang Shin-Che is a Taiwanese singer of sentimental Mandopop ballads.

A PEOPLE'S DIRECTOR FROM THE GRASSROOTS OF CHINA 15

A Conversation between Lin Xudong[1] and Jia Zhangke
Translated by Claire Huot

Lin Xudong: Is the film *Xiao Wu* a portrayal of your own experience?

Jia Zhangke: When this film was shown abroad, several journalists asked me that question. No, it's not exactly the case. Of course, the film was shot in my hometown, so it necessarily has some connection to my background when I was growing up.

Lin: Could you talk a bit about that background?

Jia: I was born in 1970 in Fenyang, Shanxi Province. My father worked in the county capital at first, but because of a problem with his family origin he was criticized and sent back to his birthplace in the countryside. There he worked as a primary school teacher of language and literature. I still have many relatives who live in the countryside. Peasant society has deeply

---

1. Lin Xudong is a documentary filmmaker, a critic, and an arts event organizer. He obtained his master's degree in 1988 from the printmaking department of the Central Academy of Fine Arts. He has served on the juries of the 1999 Yamagata International Documentary Film Festival and the 2003 Hong Kong International Film Festival.

influenced me. I willingly acknowledge and value it. It's not peasant life itself that is meaningful but the ways of surviving and of understanding things that derive from it. For instance, in the city of Beijing, how many people can claim that they have no link whatsoever with the peasantry? I'd say very few. This connection definitely influences how a person lives his or her life, in one way or another, whether interpersonal relations, values, or judgments, even though he or she physically lives in a modern metropolis. So how to properly treat that background? How to use it to empathize with the way Chinese people feel today, to examine the changes in interpersonal relations? I think that without the connection to peasantry, China's contemporary art will lose touch with the land. We need only look at the work of some young artists, which are highly subjective, narrowly private expressions.

As a kind of counterbalance to the education I received at school, I'm especially fortunate to have been in contact with ordinary people of the lower classes of society when I was very young. I was introduced to a cultural heritage deeply entrenched in Chinese folk culture. From their way of dealing with people and things, I learned how to conduct myself in the world. And in this regard, my Nanny was my greatest influence.

I call her Nanny, though normally that term is reserved for wet-nurses, and Nanny didn't nurse me. I don't know why I called her Nanny; I guess it's simply a child's expression. Nanny lived with my family in a big courtyard. In those days public order was particularly chaotic. I remember one night my parents had gone to a meeting, leaving my older sister and me alone at home. We were inside, sitting on the *kang*, when all of a sudden someone burst in and started grabbing things. We could hear it very clearly, and it scared the hell out of both of us. I don't know how children feel these days, but I remember that when I was young I was terrified by my surroundings. In those days our parents were particularly busy with work and often absent from home. They had no choice but to rely on Nanny, who lived in the courtyard, to take care of us. So, whenever there were no adults at home, my sister and I would go looking for Nanny. It was Nanny who taught us to make paper cuttings, who told us the story of Liang Shanbo and Zhu Yingtai.[2] When we were hungry, we'd eat together out of Nanny's wok.

Nanny's family was originally from Xiaoyi, the neighboring county. Her husband was a master geomancer who told fortunes. When her husband died, she moved to Fenyang with her three children. She set up a stall by the long-distance bus station and sold water for tea and boiled chicken eggs and the like. That's how she brought up three children until they were adults. Nanny esteemed cleanliness above all: she and her house were always immaculate. She was stubbornly proud. When she encountered difficulties, she didn't easily ask for help. I remember Nanny often said to me: "Be loyal to friends, treat people kindly, be filial to your parents, in times of difficulty, be brave." These were not merely abstract notions for her but absolutely practical, everyday rules of conduct rooted

---

2. Liang Shanbo and Zhu Yingtai is a Chinese legend about star-crossed lovers.

in her good-heartedness. She knew few written characters and never went to school, but I could sense her refinement. That kind of breeding does not come from books but rather from a folk tradition transmitted through generations. I believe this is real culture. I've always believed that culture is not at all the same as knowledge acquired from books. For instance, some people may indeed be well read and appear very learned, but their erudition, apart from elevating them above others, does not play a positive role in their fundamental conduct as human beings. For such people, knowledge yields cultural capital, which they see as a very useful currency. In this respect, I see more cultural dignity in the person of Nanny than in those so-called intellectuals.

That's how I lived until I was seventeen or eighteen. To be honest, though, I was a particularly confused youth; you might even say I was muddleheaded. I don't know why it was so—there are still a lot of things I haven't figured out—but I was always restless. I even went on the road as a break-dancer with a song and dance troupe, if you can believe it. You wouldn't think so to look at my fat figure now [*laughter*]. I'm planning to weave these experiences into my next film, *Platform*, which follows a touring theater group. In that film there are definitely traces of my own life back then, more so than in *Xiao Wu*.

Having said this, there are some things from that period that I'm crystal clear about: I wanted to get out; I wanted to leave that place. I didn't want to live the daily eight-hour workday lifestyle; I thought it was completely boring! I wanted to find a job in which I'd be free, not controlled by anyone. As soon as I graduated from high school, I ran to Taiyuan. At first my top priority was to find a livelihood. I'd learned how to paint a little, so I entered a fine arts class administered by Shanxi University. My plan was to get the rudiments and then do graphic design for an advertising company. My thinking then was that the profession wasn't bad, the money was good and easy. I thought I'd do that for a living, but I knew nothing about it, so I had no option but to first hurry up and study. In fact, what really attracted me in those days was writing. I was, you could say, a "literary youth."

Lin: When did you start writing?

Jia: When I was in fifth grade in primary school, the *Shanxi Youth* published an essay of mine on Taiyuan's Jinci Temple. When I was sixteen or seventeen, I started to write novels. By the time I left Shanxi, *Shanxi Literature* had already published a novel of mine. In those days, I was esteemed by the Shanxi Writers' Association; they sought me out for a chat. They said, "We'll soon have a literary college here," and they encouraged me to come. "We'll give you a salary, you can write novels." I'd already made my mark in Shanxi.

Lin: Did this recognition have any kind of concrete impact on your situation?

Jia: It was mainly an emotional satisfaction.

Lin: How was your actual life then?

Jia: In those days, I lived with a couple of painter friends in Xuxi, in the southern outskirts of Taiyuan. Our place was close to the railroad; our neighbors were peasants: small fruit vendors, long-distance transportation truck drivers, and the like.

In the beginning, I had some money I'd brought from home, but I soon spent most of it. I started going out to find odd jobs. My classes at Shanxi University were only half-days, so my friends and I would paint screen partitions in private homes or restaurant signs—that kind of thing.

The experiences I underwent in those days are typical of any person in China moving from a small town to a city to make a living… and that includes being roughly woken up from a deep sleep in the middle of the night, ordered to get up, and interrogated. That's when you know in your bones what your true status is in that city: you don't have a resident's permit, you don't belong to a work unit with a regular steady job. For some people, you're what they call tramps. Compared to those who had regular jobs, we suffered a great deal more and worked much harder, but in those days, on this issue, we had no right of expression whatsoever. I felt the injustice deeply.

Those were the living conditions that gradually shaped my attitude toward life: I would not put my trust blindly in anybody, anything, or any group; I believed that it is only through your own efforts that you will reach your goals and prove the value of your own existence.

Lin: When did you decide you wanted to make films?

Jia: In 1990, exactly when I turned twenty—I think it's time I answered this question honestly and admit that it was after watching *Yellow Earth*.

The film had been out for quite a while, but I'd never seen it. That day, by chance, the movie was being shown in a Taiyuan movie theater not far from where I lived. I went to see it. This was also my first time seeing a film made by the Fifth Generation. After the movie, I felt overwhelmed by film as a medium. I realized that it was totally unlike all of the ways of expression I knew, that it possessed so much more scope and potential. In addition to the visual component, there was the aural, and the way it represented temporality—in that relatively short duration, you could convey such an abundance of life's experiences. I thought, "I must make movies!" and rushed off to Beijing to take the Film Academy's entrance exam.

Lin: In other words, it really was that one film that changed your...

Jia: That's right! At that time it absolutely changed the course of my life. If it hadn't been for that film, I probably would be making money in a Shanxi advertising company, maybe I would even have created my own advertising company, because my highest goal has always been to own my own business, to be my own boss.

Lin: Looking back, what do you now think of the film *Yellow Earth*?

Jia: Recently I've seen it again, and I was still very moved. The truth is I can't deal objectively with that film.

Lin: It carries too much emotional baggage?

Jia: You could say so. First, the way it portrays the Yellow River region, the Loess Plateau… for me that in itself is replete with emotions—I'm from Shanxi, I was raised in that area. Second, the film itself is part of my own personal experience as a youth, because I'm very aware that it's what made me change paths. So I can't treat it rationally. When I view it, I immediately recall that time of my life, 1990, 1991…. struggling to study painting, striving to make a living… And afterward, everything suddenly changed.

Lin: You said you had special feelings for the Yellow River and the Loess Plateau, but I don't see any images of that scenery in your film.

Jia: You're right, there aren't any. For me, that way of life is gone. My life now is what it is, and that's the way I need to show it on screen—there's no point in adding other stuff.

Lin: When did you enter the Beijing Film Academy?

Jia: In 1993. I was older than most people in my class.

Lin: How did that feel?

Jia: I thought they were all kids, unlike me who had lived out in the real world. Especially since I had already earned money. I had brought money with me that I'd earned myself.

Lin: How long were you able to live with that money in Beijing?

Jia: Not very long.

Lin: So how did you solve the problem of tuition fees and living expenses when you were studying?

Jia: My family chipped in and I sometimes did odd jobs. I was a "sharpshooter," a ghostwriter who helped people write some very boring television scripts. You get paid, but you don't sign your name.

Lin: What specialization did you study at the Film Academy?

Jia: I studied film theory in the literature department. At first, I wanted to try for the directing department, but the competition was too high; I was afraid I wouldn't get in, although I always wanted to be a director.

Lin: What do you feel is the most important thing you got out of the Film Academy?

Jia: The opportunity to learn film history in a relatively systematic way; that way you don't reinvent the wheel.

Lin: Under what circumstance did you start to create *Xiao Wu*?

Jia: In 1996 I went to Hong Kong with my fifty-minute video *Xiao Shan Goes Home* to participate in an independent film competition. During that event I met an investor. He was a graduate of Université Paris 8, where he'd studied film theory. When he returned to Hong Kong he put together a small company called Hu Tong Communication. He really liked *Xiao Shan Goes Home*. He asked me, "How much did you spend to shoot that thing?" I said, "Well, maybe a few tens of thousands." He said, "Oh! With a few tens of thousands one can make a film? Let's make one together." He told me, "Right now, I don't have a lot of money, but I'm getting ready to invest in film." He added, "Let's start small: spend a few hundred thousand to make a short and put it out on the circuit. Once the company grows, we'll do a larger-scale film."

So I came back home and wrote a script for a thirty-minute film titled *Tender Is the Night*, like Fitzgerald's novel. It's about a young couple's first romantic meeting at night and takes place entirely in a closed setting, within a short span of time. We were planning to do an experimental short film.

Once the script was done, cameraman Yu Lik-wai came over from Hong Kong, as we'd initially agreed to do the film together. But it was already close to the Spring Festival, so we planned to start shooting after the festival. In the meantime I took him to my hometown.

I hadn't returned to Fenyang in about a year and a half. When I got back this time, I found that huge changes had taken place.

Every day during the Spring Festival, classmates and childhood friends dropped by to chat. During these conversations, I realized that everyone seemed to be living in some sort of predicament. I couldn't figure out how, but every person had run into trouble in their personal relationships: husbands and wives, older and younger brothers, parents and children; or with neighbors—a whole gamut of real conflicts of interest had strained the relations between people of this small county town. Among them were a few close friends from my childhood, people I'd grown up with. By the time they'd turned eighteen, their lives had come to a halt; there was nothing left to look forward to; they got into a work unit, found a factory job, and then they were trapped in the cycle of everyday life. The dejection, the loss of any romance in human relationships…it was shocking.

Walking in town afterward, my astonishment only grew. At the edge of my old hometown there used to be an "open economic zone" called Fenyang Bazaar. Clothes and such things used to be on sale there. But when I went to take a look this time, all there was were karaoke bars! Everywhere, female escorts from the North East and Sichuan walked the streets. People's jokes and conversations veered to that topic too. Another shock was the main street in Fenyang, which we locals call Center Street. It's not very long, no more than a ten-minute walk from one end to the other, but both sides of the street are lined with old buildings occupied by various types of shops. Someone told me, "Next time you come, you won't see any of this. In a few months all of these buildings will be torn down to make way for new high-rises. Fenyang is being upgraded from county-town to county-city." Today that street is one homogenous row of new constructions, all in glazed brick. The news of the imminent demolition of the buildings was also a factor that inspired me to make *Xiao Wu*. It's not about nostalgia for the old, but by showing the course of the demolition, you can visualize the profound and concrete effects of social transformations on the lives of people of the lower strata of a small town. I saw in that impending situation a chance to write something serious. My creative drive went into full gear. It was exhilarating!

I was excited because after a few years in the Film Academy, all the Chinese films I'd seen could roughly be put into two types: either commercial and consumer-oriented or ideological films. Films that record honestly the changes of the decade are far too rare! The entire nation is in a critical period of transition, yet nobody, or at least very few people are doing this work. I believe that people in our profession find this state of affairs disgraceful. If only as a question of conscience, it seems to me that we ought to find ways to make films that honestly reflect today's climate.

That's how I scrapped my original plan and began to devise a new script. At first I wanted to write about a craftsman, maybe a tailor or a blacksmith. In any case, it would be a person who tries to earn a living by doing traditional craftwork on some small town's main street. In our day, such a lifestyle meets obstacles of all sorts coming from every direction, yet this obsolete role is the craftsman's only way to participate in society. From this specific case, I wanted to depict how people can be worn down and broken spiritually. When there are dramatic changes in the course of history, things previously believed to be unalterable also change. Some people fail to adapt and suffer greatly.

While I was writing the script, friends often dropped by to chat. Among them there was someone working for public security. We'd been in the same class from primary to high school. One day he asked me, "Do you remember Donkey from our class? He's turned into a thief and is in jail now. I'm the one guarding him. He often steps up to the bars to talk with me. We talk philosophy...." When I heard this, I became really interested. Suddenly I thought I might create a kind of hybrid: a craftsman-thief.

Sometime later, someone said to me, "Your choice of a thief as your main character lacks ordinariness, it doesn't fit with your creative design to record this era." But it seems to me that the universality of a fictional character cannot be judged in terms of its specific social status. The point is to what extent we grasp this specific character's humanity. I'm interested in the character of a thief because it provides an interesting angle to show the transformation of social relations. For example Xiao Wu's friend Xiao Yong was formerly also a thief, but by trafficking smuggled cigarettes and opening karaoke bars, he was transformed into a respected private entrepreneur. And this, in turn, changes societal values: the trafficking of smuggled cigarettes becomes a trade in its own right; the opening of karaoke bars, an entertainment business. A person like Xiao Yong is in his element in that world, he easily alters his position in society. Not so for Xiao Wu: wherever he goes, he's always just a thief.

Later on I realized that, subconsciously, something else had triggered my choice of such a character: Vittorio De Sica and Robert Bresson, two directors who have had the greatest influence on me. Both feature people who steal in their work. *Bicycle Thieves* and *Pickpocket* were my favorite films when I was studying. But when I was writing the script I wasn't at all aware of their influence.

Lin: A while back you mentioned "serious writing." Does that mean that for this film's script you chose to use a relatively traditional writing method?

Jia: Yes. Because I felt that it was only by doing so that I could fully develop all the various levels of this film's narrative. I didn't want to treat a topic as critical as *Xiao Wu* in an abbreviated fashion. And, faced with the intricacies of life today, I wanted to create a narrative framework with a clear and ordered structure to pave the way for the actual filming.

During the writing process, I did my utmost to diligently work out the links between the various dramatic elements. I had many moments of hesitation and did a lot of revision during that period. I was continuously discovering and adding new elements. In one sense, the process allowed me to pour my heart out, but at the same time, it tested my ability to restrain myself. I kept exhorting myself to be vigilant during the writing process, to keep my emotions in check.

Self-restraint was essential in this phase and for this particular film. By then, I had already figured out that my direction would incorporate improvisation, interaction, and a semi-documentary style. I felt that this method would allow me to create the film I wanted to make. Having made that decision, I was acutely aware that the filming would be an adventure in shooting on location. And I knew from experience that when you're shooting in a natural setting, unexpected things happen, which can yield all kinds of possibilities. But when it comes to the choices available to achieve your purpose, these are in fact quite limited. As a result, in the midst of countless fortuitous elements, you must have a firm grasp of your story if you hope to maintain the direction you want from start to finish. Without such clarity, what seems acceptable on location turns out to be a pile of useless trash in the editing room. The first time I took hold of a camera to make a film, that much was clear to me. But that grasp of the story line compensated for my lack of experience and allowed me to stick to a tight budget and to approach my aesthetic goal. Under such constraints, I knew that unchecked petit-bourgeois sentimentality would only lead me astray.

That's the way it was during the scriptwriting phase. I prepared as exhaustively as I could. At the end of the fifteen days of the Spring Festival, I returned to Beijing with a completed script. I immediately sent it to my investor, and he was very excited by what he read. After figuring out a few operational details, we started to film.

Lin: How long did the shooting last?

Jia: Twenty-one days of real shooting.

Lin: How much money was invested?

Jia: In the early stage, approximately RMB ¥200,000 [approximately USD $40,000], which was mostly spent to buy film and pay for equipment rental. The film crew basically received no remuneration. If you add the expenses in the later stages, the total investment was RMB ¥380,000 [USD $61,000], give or take.

Lin: In *Xiao Wu* you used non-professional actors exclusively. Was it because of the tight budget? Or was it an aesthetic decision?

Jia: It wasn't entirely because of the money problem. I have many friends who study acting; if I need them, I know that they'll help out without getting paid. The reason why I used non-professional actors is because of the style of film I wanted to make. For the Hong Kong publicity poster, I wrote: "This is a film in the rough." In other words, my film contains no smoothing of the edges or contrived artifice.

Lin: Could we say that this "roughness" is your own view of real life?

Jia: It's an attitude; it comes from my direct experience of life in the lower strata. It's not because such living lacks glamor that it can't be faced squarely. The same goes for the characters in the film: I want to convey their humanity in their particular living circumstances. Their existence is unsophisticated—coarse but full of vitality, like weeds along a road. That's why I want the actors in my films to be spontaneous, to allow unconscious things to materialize. For example, I chose Wang Hongwei to play the main role because, even in front of a camera, he's able to maintain the natural traits that make him who he is in normal circumstances. This is what makes him appealing. His body language is particularly vibrant; a professional actor could not easily attain such a dynamic liveliness. Professional actors generally undergo physical training. Their basic body language is not natural; it's programmed and results in a conscious and artificial demeanor.

Lin: But after all, you're directing these people to act in a fictional world. How do you go about getting the performance you want from non-professional actors?

Jia: To achieve this, everyone has different techniques. For example, some people will film and film, shooting twenty or thirty takes for one scene. Sometimes even up to eighty or ninety takes. I don't think that's a good method. In any case, my budget really doesn't allow me to shoot that much footage. I have to limit my overall shooting ratio to about 3:1.

For me, the first step is to assure that the script is solid, that the relations between characters, plot development, and rhythm between scenes are all structurally sound so that I have a logical framework in my head. That way, on location, there won't be any disorientation and confusion when unexpected elements arise. As I said earlier, the more methodical you've been in the early phase, the greater room for on-site improvisation. For example, in the scenes with dialogue, I didn't write the lines down. On set, I explain clearly to the actors the gist of the plot element and the basic performance required. Then, when it's time to actually shoot, it's up to them to improvise the lines the way they understand the situation. Their acting is very relaxed. The outcome is great most of the time, even occasionally surprisingly fantastic. But the prerequisite for this technique to work is that you must fully understand the significance of that particular episode in the context of the entire film.

When working with non-professional actors, a director has to find ways to help the actors dispel their stage fright. He has to create favorable conditions on the set. For instance, by using high-speed film, there's no need to add lighting, or at least very little. Non-professional actors are often afraid that once they enter a scene they won't know what to do. They stand there stiffly. So, before you begin the scene, you have to help them find a support—a small accessory to hold or a small gesture can help. Once they get the feel of it, their delivery will develop its own unhindered rhythm.

All this said, the most vital thing, in my opinion, is to establish a climate of trust with the actors before the filming starts. Everyone, not only the director but also the entire production crew, including the script and set workers, needs to be complicit with the actors. If their environment is not unfamiliar, they'll totally trust you and the camera and feel at ease to perform. As a director, you sometimes have to spend a lot of energy to coordinate all kinds of interpersonal relationships within the team.

On the whole, the actors I found for *Xiao Wu* are people I'm pretty familiar with and understand well. But I've also experimented with actors in a somewhat impromptu fashion—choosing extras intuitively on location. Before they fully realize what they're doing, I get them to act. Strictly speaking, you can't say that they're acting; they simply follow your instructions and make the motions as they would in real life. For example, I found Xiao Wu's parents in the village two to three hours before shooting the scene from the crowd of onlookers. When I made that decision, there was quite a bit of resistance. The cameraman argued with me: "It won't work, we only have a total of forty rolls of film; what if we need more?" We were using Kodak Vision-500T high-speed film, which is rare in China. Were we to use it all we'd have to find a way to have some sent up from Hong Kong. The pressure was indeed enormous, but in the end the outcome was pretty decent. In my experience, if you use this method, you have to keep a tight control of the schedule and pace: you ought to finish the shoot in that one day because, if it takes longer, the extras will start turning it over in their minds: "Wow, I'm in a film!" And then they'll start "acting;" they'll "act" according to the "acting" they see every day on television. And then the whole thing is ruined.

The process of connecting with actors is hard to explain. I can only say that I might be able to offer a few pointers from my experience but no set formula, because it's different in each case. Actually, it depends on how well a director knows human beings. It's a very delicate thing to understand a person's psychology, their self-image, their state of mind, and even harder to be able to infer experiences they may have been through. Armed with this knowledge, when you meet with the untrained actor to explain the scene, there are a few tricks to work with. For example, if you tell some people to go east, they'll go west. If you reverse it and tell them to go west, they'll move in the right direction.

Lin: Could you explain that?

Jia: For example when we shot the public bath scene. That was a test for me—it's a crucial scene. When I first told Wang Hongwei he'd have to do this scene in the nude, he said fine. But before the actual shooting, some crewmembers were a little worried; they kept telling me, "You have to talk this through with him, so that he doesn't quit the scene in disgust at the last minute." But I knew this guy. I knew I absolutely shouldn't say more; I should just trust him, let him feel he was trusted. That was the only way to pump up his courage. To discuss it at length would have upset him, and he might have just quit on the spot. This relation is really singular. That's why, before shooting, I didn't talk to him about it and didn't allow the rest of the crew to bring it up—until the very morning of the day of the shoot, not a peep. And when it was time, it happened naturally, and we shot that scene in one take.

Lin: People who've seen *Xiao Wu* all seem to have been deeply impressed by Wang Hongwei's performance. In my experience, non-professional actors can play their part brilliantly; but Wang Hongwei's acting is not only great, his performance commands the entire film. I'm not sure how, in practical terms, you succeeded in such fine-tuning with him? Is it because in life you know this actor very well, and so, while you were working out the story, you half-consciously used this person's quirks to build the character? Or did you consciously design the character according to the actor's particular traits in his everyday life?

Jia: Let's say, we know each other too well. I often observe his gestures, such as his quivering sleeves—when he talks to me he likes to shake his sleeves; it's pretty funny. Such things have helped me to perfect my idea of the mise-en-scène and plot elements.

Lin: Could I put it this way: these impressions you gather from daily life will at some point spur your creative inspiration? In other words, offer you fodder to work with?

Jia: These things provide me with workable elements specifically for that particular person: the way he spits, reacts, walks....

Lin: Did you and Wang Hongwei grow up together?

Jia: No. He's from Anyang, Henan. He played Xiao Shan in my first video work, *Xiao Shan Goes Home*. Xiao Shan is a youth from the country that goes to the city to find work. He's an introvert; he doesn't like to talk much about things. He prefers tacit communication.

Lin: Will Wang play the lead role in your next film?

Jia: That film will have four main protagonists; he'll be one of them. He'll be the boss who takes the troupe on tour.

Lin: Aren't you worried that his performance in the two films will be too similar?

Jia: I do worry a little. Before the filming, I'll find some ways to solve that question for the better. For instance, in terms of the character's expression, in the next film I'd like him to play a character that would have more comic sides, which would lighten him up, so he won't be so heavy.

Lin: In *Xiao Wu*, the main character doesn't know how to sing or dance. Did you base this on the actor's real life?

Jia: He really doesn't know how to dance, but he loves to sing. It's set up that way in the film particularly for dramatic purposes. It shows how the protagonist, in contrast to others, can't adapt to reality; he's uneasy. Female escorts working in karaoke bars, and karaoke culture itself, are specific signs of Chinese society in transition. Karaoke culture reveals a great deal about overall contemporary Chinese society, history, and culture. Hu Meimei does this peculiar escort job, but she's also a woman, a pretty woman in the eyes of any man. Xiao Wu, for his part, may seem worldly, but in fact he has no experience with women. He's practically clueless, and his cool facade can't hide his bashfulness. Though he too frequents the singing parlor to flirt with the girls, deep inside he's very traditional. He's shy about expressing his own feelings. He sings but only alone in the bath and for himself. I placed this particular scene in the golden mean of the story. I wanted to get inside the character when he's at his truest, to achieve a relatively smooth transition in the narrative, and thereafter to move the story toward its climax. In adopting this structure, I followed a rather traditional, classic convention.

Lin: Are there any major modifications of the original script in the final version of the film of *Xiao Wu*?

Jia: There are some changes, but they're mainly adjustments of scene segments and rearrangements of some specific details. Because I wrote the script quite meticulously, I knew where I was going, so that on the set I could open all my senses to discover new things in real time and space. For example, the script ended with the policeman traversing various places with Xiao Wu in tow. It was also an open ending, but I felt it was sort of dull. Later, filming on location, there were always lots of onlookers we couldn't chase away, so we tried to film around them as best we could. It slowly dawned on me, why not make a scene out of these bystanders and incorporate it organically into the film? That's how the present ending came to be. I find that, compared to the original script, it's much more appropriate and far more expressive.

Lin: There's also that scene of karaoke singing on the street that's deeply moving.

Jia: We chanced upon that scene at the end of one day when we'd finished shooting and were on our way home. We were on a bus and saw a crowd of people singing karaoke in front of a shop selling funeral wreaths. It was like being in a dream. The artistic director, Liang Jingdong, told me our film could use such a scene. I agreed, so we later devised a similar sequence.

Lin: Many popular songs are featured in *Xiao Wu*. They infuse the film with a strange feeling of anguish—human sorrows and struggles bizarrely packaged by popular culture.

Jia: Yes, many pop songs like "Rain in My Heart," "The Warrior Prefers the Land to the Beauty," and Tu Honggang's "Farewell My Concubine,"[3] along with the

---

3. "Rain in My Heart" ("Xin yu") was performed by Taiwanese singer Lily Lee. "The

street noise of bicycles, trucks, motorcycles, especially motorbikes—all sounds I heard all day long in the small county town of Fenyang. My original plan was for the film to document the times, not only in terms of visualizing the spring of 1997 in a small town in Northern China but also to include a record of the sounds of that time. So, to match the narrative, I purposely chose that year's most popular and representative songs of the karaoke culture, or you might say the most common ones, especially "Rain in My Heart," which every karaoke-goer was singing. It created a strange sense of belonging. Such things are, in fact, a reflection of society's morale. For instance, I wanted to convey on film the depressed machismo of "The Warrior Prefers the Land to the Beauty" or the fragile bravado of "Farewell My Concubine." The combination of these songs with the images helped to set the tone for the entire film.

Lin: You've talked about the archival aspect. I've also noticed that your film has a raw feel straight out of life. Did you purposely want to give this real-life effect to your film?

Jia: Yes. Some people think that it's easy to do this: you simply pick up a camera, point, and shoot. That's because they haven't tried it. In fact, the experts in the field know that it's far from simple. You have to have a lot of experience and understanding of what shooting on location

---

Warrior Prefers the Land to the Beauty" ("Jiang shan mei ren") was performed by Jay Chou, who is also a Taiwanese singer. Tu Honggang, who sings "Farewell My Concubine" ("Bawang bie ji"), is a Mainland singer.

means. This issue—what makes a "good" film—is one on which I disagreed many times with my classmates. For example, some of them really liked *Legends of the Fall*—that sort of thing. Of course, everyone is entitled to his or her own criteria. I don't judge a film by how pretty the lighting is or how intricate the camera movements are. The most important thing is whether the film expresses the stuff of real life, whether it offers insights into reality. But the truth is that people mostly don't want to see films that have a real edge. It makes them uneasy, maybe because such films require viewers to face reality head-on, and it seems most people don't want to endure such things in film. They'd rather consume finely polished and glossy products.

Lin: That's also a reality. Today, cinema is an art that must take into account financial profits. With art now operating as an industry, are you confident that you can keep on working the way you want to?

Jia: Now that I've made *Xiao Wu*, I think that I can get by, bit by bit, with a relatively small budget. What's vital is to have faith in what you're doing.

Lin: About the images of *Xiao Wu*, it may be because, for a number of years now, I've been making documentaries, but I'm deeply impressed by cameraman Yu Likwai's documentary-style cinematography. Not only did he astutely bring into play all sorts of impromptu elements, he also never lost his grasp of the film's content and never showed off his technical virtuosity.

Jia: Our collaboration was a very happy experience. Yu Lik-wai and I met at the Independent Film Festival in Hong Kong. He grew up in Hong Kong, and he's four years older than I am. He studied advertising photography in France. Then he went to Belgium, to the film school Institut national supérieur des arts du spectacle in Brussels for four years. At first, some members of the team had doubts. They felt that it would be hard to communicate with a person from Hong Kong, someone with a very different background, on a film that reflects the life of the lower strata of the Mainland. But actually, Yu Lik-wai's circumstances are not that straightforward. His parents were leftists in Hong Kong, so from an early age he read a lot of publications from the Mainland—*China Pictorial* and the like. Later, he made two documentaries in Beijing.[4] He understands the situation here very well. In any case, most importantly, our conceptions of cinema are quite close: we both love Bresson's films; we share a common language.

Lin: During the filming process itself, how did you work together?

Jia: I don't look at the monitor; I never do. I stay on the set and arrange the mise-en-scène. I give instructions on the gist of the scene. As for the specifics of framing, composition, lighting, and the rest, I leave that to Yu to figure out. I let him do it. These days everyone likes

---

4. The only known documentary that Yu Lik-wai made in Beijing is *Neon Goddesses* (*Meili de hunpo*) (1996), which follows three sex workers who migrated to Beijing.

to use the monitor. I've used it in the past, but now I don't think it's a good habit, at least not for me. If you only use the screen to control things, you'll ignore a lot of interesting things going on in the location. You'll end up picking up the seeds and discarding the watermelon. On set I prefer to use my eyes to make decisions. That way you argue less with the cinematographer. Making a film is a job with many responsibilities, and since you chose this cinematographer, you should fully trust him; keep your hands off and give him free rein. Otherwise, you should definitely find someone else.

Lin: You've mentioned Bresson a few times. And earlier you said that *Xiao Wu* has some sort of unconscious link with Bresson's and De Sica's works.

Jia: I discovered that when I was giving talks in Europe. Once, after viewing *Xiao Wu*, someone in the audience asked me, "Whose films do you like?" I answered, there's Bresson and De Sica. That's when I became aware that, between their film methods and my creation, there might very well be some underlying connection.

What first attracted me to De Sica's films was his empathy. It's the most basic thing—how you approach life. Just as important is that he showed things in a very cinematic way by structuring the film through images that flow one into another. From his films, I learned how to find ways within the film medium to discern and express the beauty in very down-to-earth, real-life situations. Behind his films' naturalistic style that shows things as if randomly selected lies a meticulous ordering of things. He didn't make it up or have people put these things together; he dug them out of mundane life.

For example, *Bicycle Thieves*, in terms of narrative, is composed of the following plot elements: bicycle — work — lost bicycle — lost job — looking for bicycle — unable to find bicycle — steal a bicycle. The plot's development is accompanied by a visual composition, for example: dawn — morning — noon — afternoon — dusk. These are variations in time, but there are also variations in the weather: windy — rain — blazing sun of high noon. He combines these elements organically with the story, creating a flowing structure of images. During my studies, it was in his films that I discovered the aesthetics of realism. I began to understand that there is no real barrier between pure realism and expressionistic or surrealistic content. As long as you understand them well, it is possible to freely go back and forth from one camp to the other. This is what I strove to achieve in *Xiao Wu*. I believe that De Sica's contribution to cinema, apart from his opening up cinema to social issues, is his creation of a fascinating film aesthetic.

On this point, Bresson attained an even higher degree of refinement. The year I went to Hong Kong, I saw his *Pickpocket* for the first time. I was dumbstruck. Using what seems like mere sketching, he drew the outlines of a very real material world. But pulsating in the background of such a world, you can sense that there is something entirely formal, very intelligent—it's a positioning arising from attention to the small things of ordinary life.

Maybe my aesthetic preference comes mainly from my experience of reading Borges's novels. Of course I've read him in Chinese translation, and I can't judge his original style. In the translated works, I encountered a series of concrete images unadorned by rhetorical flourishes. Borges uses such concision to construct a perplexing imaginary world. This is precisely what I want to accomplish when I make a film. For example, in *Xiao Wu*, after the sequence where Meimei kisses Xiao Wu, music from John Woo's *The Killer* is heard off-screen. It creates that distancing effect that allows us to oscillate freely between the two planes of reality and non-reality.

Lin: You also said earlier that *Yellow Earth* was a key influence. I think you probably saw quite a few films by Fifth Generation directors after that. How do you feel about them today?

Jia: I find that their films can be divided into two phases: before and after they achieved success. The differences between the works in these two phases are enormous, especially for Chen Kaige. In his early films, we see a side of him that's very earnest, marked by many courageous risks. Even though we might argue about his film technique, such as his emphasis on visual plasticity and the like, at least for Chinese cinema of the time, he had a positive effect. Later, he basically turned commercial—the stuff he's done like *Farewell*

My Concubine (*Bawang bie ji*). That type of popular drama—at best we can only say that it's a highbrow commercial film. And then there's Zhang Yimou.

Lin: How do you understand the turnabout in their creative work?

Jia: For me, what happened to them serves as a lesson. Maybe at first they didn't think too much about it; it just happened because of all kinds of pressures and temptations, because of ideology or investors. Unwittingly, little by little, they changed. I'd also say this change is symptomatic of a loss of faith in the power of cinema.

Lin: So where do you stand now?

Jia: I believe I'm a people's director from the grassroots of China.

Lin: Will you always remain that way?

Jia: These past few months I've been thinking about this question a great deal. At the Berlin Film Festival, the president of the Youth Forum, Ulrich Gregor, asked me that question after giving me a prize. It's true, success has given me some means and has also brought some temptations. For example, now that I have the award money, I can take a cheap cab, I don't need to squeeze myself into a bus anymore. In my present circumstances, my state of mind is already different: if

I were to keep on taking crowded buses, I wouldn't experience it the same way; it would be posturing. In other words, in the course of your constant pursuit of success, unwittingly, little by little, you lose what's fundamental to you. When I think about this, sometimes I'm really terrified.

Lin: What do you think of the young directors in your own generation? I mean the films made by the directors called the Sixth Generation or the New Generation?

Jia: It's not easy for me to answer that because my emotional and intellectual response has also changed. Before I made films, I was more prone to notice the flaws in their filming. Later, once I started to make films myself, I slowly realized the significance of their work, which does not lie in their technique but in the great efforts they made in this particular climate. During the maturation of the Fifth Generation, filmmakers basically still operated within the industry that was based on the planned economy system. Concomitantly, they drew strength from the "emancipation of thinking" ideology. However, when the next generation of filmmakers began to make films, they had to produce scripts under ideological pressure and had to find funding and even find ways to promote their own films. In other words, it was only with the Sixth Generation that independent filmmakers emerged and had to face serious problems. That they were able to produce any work at all within the tiny opening of the economy and ideology and with the cultural industry climate of that time is a cultural feat in itself—even though their films had many flaws. That's why I don't think it's relevant to discuss any more the quality of this or that particular work.

West Tayuan, Beijing, June 1998.
Originally published in the journal *Today* (*Jintian*), volume 3, 1999.

IMAGES TAKEN FROM THE WORLD OF EXPERIENCE   16

Interview by Correspondence between Sun Jianmin[1] and Jia Zhangke
Translated by Claire Huot

EXPERIENCES OF LIFE AND EXPERIENCE IN FILM

Sun Jianmin: It seems that, for filmmakers, early experiences, especially the teenage years, are critically influential. Many great masters of cinema have even made movies that are based directly on their personal experiences; for example, *The 400 Blows* and *Cinema Paradiso*. The last time I interviewed you, for Volume 7 of *Avant-garde Today* (*Jinri xianfeng*), you spoke extensively about your film-viewing experiences during your youth. This time, could you speak more at length, especially about your personal experiences in that small county town in the hinterland of China? It seems to me that if you hadn't undergone these experiences in your youth, *Xiao Wu* or *Platform* would not exist today.

Jia Zhangke: I was born in May 1970 in Fenyang County, Shanxi Province. My father was a middle school language and

---

1. Sun Jianmin, a graduate from the literature department of the Beijing Film Academy, has collaborated with Jia on several films and in many capacities. Sun is a novelist, screenwriter, and film director. He teaches at the Shanghai Drama School.

literature teacher and my mother was a sales clerk in the retail department of the county's Sugar Industry Tobacco and Liquor Company. I also had a sister six years older than me. She was a member of the school's propaganda team[2] and it meant that she often missed school, because there were so many performances. Their most representative work was "Train Heading to Shaoshan."[3]

The first time I saw my sister performing on stage was in our primary school's playground. I don't remember what the mass rally was about that day, but a neighbor secretly took me to see the show. From some distance I heard a familiar tune being played on a violin. I squeezed into the crowd. My sister's face was painted, and she wore a flimsy outfit as she performed in the cold wind. Her performance didn't leave me with much of an impression, but to this day, I can't forget the dense mass of people packed together in neat rows around the stage. I don't know what force drove so many people to gather there to listen so meekly to someone's voice over the loudspeakers, but that scene gave my child's heart its first surreal moment, a feeling I nurtured until I was twenty-nine years old. I had to reproduce that on film. It became the prologue of *Platform*. And since then, loudspeakers recur in all my movies. It's become a sound in my life that I can't let go of.

Our county hub wasn't big. You could ride your bicycle straight through, from east to west, or south

---

2. A propaganda team would perform for workers.

3. A revolutionary song from the beginning of the Cultural Revolution. Shaoshan, in Hunan Province, is Mao Zedong's hometown and a symbolic rallying place.

to north, in less than five minutes. Surrounded by mountains and rice fields, the town itself was made up of old buildings in dark flashed brick. Apparently this county town existed as far back as the Qin dynasty (221–206 bc). Located on the Loess Plateau, to the east are Pingyao and Qixian. The Shanxi merchants used to do brisk business in that area. Westward, soldiers crossed the Yellow River to reach as far as northern Shaanxi. Yet, Fenyang still suffers frequent blackouts today. In my youth, as of nine o'clock in the evening, it was lights out.

I used to ride my bicycle all day long through my tiny county town. I was like a ghost in the legends, pounding the walls, pacing back and forth, spinning around in every direction. The movie house was showing a new film, but the projections stopped for the Three-Level Cadre Meeting (*san gan hui*),[4] which was being held in the theater. The magazine *Popular Movies* (*Dazhong dianying*) had just appeared on the newsstand in front of the post office. On the front cover was the actress Gong Xue,[5] but if you didn't buy the magazine, you couldn't leaf through it. The usual people came and went but not a single fresh face among them. There was really no place to go in town and never anything to do. Fortunately, often there was news of a fight around the bus station. A bike would come through carrying a guy with a bloodied face heading toward the workers' and office staff's hospital. You'd want to follow, but then you realized that

---
4. A general meeting of the cadres from the three county-level divisions: city, town, village.
5. Gong Xue (b. 1953), was a rising-star actress in the 1970s.

your classmate's older brother had just borrowed your bicycle. During sandstorms, we would lean against the back wall of someone's home and smoke a Princess cigarette in secret while we watched the wind rattle the electric wire. From the military barracks in the distance you could hear the radio rebroadcasting the *News and Newspaper Summary* program. The cigarette wasn't quite finished, but it was time to go home.

At school, my grades were never good, but I had lots of friends—a dozen or so sworn brothers in primary school. By junior high, half my friends had quit school. They didn't have jobs; they just hung out in the streets. My father forced me to study. My friends would wait every day outside my school for classes to be over, and then a gang of us would bum around town.

Around that time, video playhouses first appeared. I watched countless martial arts movies from Hong Kong, including King Hu's (Hu Jinquan) *Dragon Gate Inn* (*Long men ke zhan*) and *Raining in the Mountain* (*Kong shan ling yu*). Since there was no outlet to release energy, when we weren't at the video playhouse, we'd get into fights. Graduation from junior high marked another milestone: many of my remaining classmates quit school then, some to become soldiers, most of them to hit the streets.

Once I went with a friend to see a movie; after buying his ticket he said he had to go to the washroom, so I went on ahead inside. I looked everywhere for him but, once I left the theater, someone told me he had been arrested. On an impulse, he'd stolen a woman's watch. At that moment, I realized that life can suddenly throw huge misfortunes onto your path. I had a vague feeling that my character was changing. Shortly after, a friend working at the Xinghuacun Fen Wine Factory thirty *li* [over nine miles] away died on his way back to town after drinking too much. And later, during the "crackdown on crime" (*yanda*) campaign,[6] friends were constantly being arrested and jailed. I became depressed; it seemed that life just unfolds according to an inexorable fate. I wasn't prepared for that; I was at a loss. At that moment I realized I had already become an adult. I knew I had to leave.

Fenyang has no railroad; trains don't go through there. So the first thing I did when I learned to ride a bicycle as a freshman in junior high was to get my classmates together and secretly ride our bikes to another county town called Xiaoyi, three to four *li* [about one mile] away, to watch the trains. We kept looking until finally we spotted the railroad tracks. We all sat on the ground, holding our breath to hear the faraway sound. It was like a ceremony; it made us feel like there were still awe-inspiring things in life. Finally, a slow coal-carrying train rumbled by, its sound fading away little by little, becoming a kind of summons. To us the railroad signified the future, faraway places, hope.

In the beginning I had no idea that, in my heart of hearts, I harbored an intense desire to tell these stories. Once I had left the county town for good, those people and events from Fenyang became gradually clearer. Inwardly, I was anxious as I felt my unshaped inner self yearning to express itself.

---

6. The first of the four "crackdown on crime" campaigns took place in 1983.

Sun: After leaving that county town, you arrived in Beijing, the big capital, "the international metropolis," to study at the Film Academy. From what I know, in China's art schools, there are always a lot of children coming from influential artistic families. Even before enrolling, many of these people already live in that world and have probably already had hands-on experience. They also seem to have a reserve of professional knowledge; they appear more authoritative. As someone who lacked both city life experience and an insider's knowledge, did you find this environment intimidating? Did this influence your own thinking, creativity, and film practice? Did it indirectly make you the director you are today, whose films are different from all the Chinese directors of your generation?

Jia: When I arrived in Beijing in 1993, the Third Ring road wasn't finished; migrant construction workers lived all around the Film Academy. I was very familiar with their ways and expressions. When I walked around in the streets, I didn't feel all that far from my former life. But in the Film Academy, when students attacked one another, they would call their target a "peasant." I was always somewhat taken aback not only because I had a strong peasant background but also because they manifested such a lack of upbringing. That's why every time someone said that the Film Academy was a college for aristocrats, I would laugh inwardly: How can aristocrats have such a lack of breeding? They didn't even have a veneer of respect. It was then that I discovered the value of my own experiences and upbringing.

IMAGES TAKEN FROM THE WORLD OF EXPERIENCE

I realized that my social background has been despised by screenwriters in China. Where I come from is a world that the established film industry, with its sense of superiority, does not bother to explore. It seems that Chinese film directors don't want to face their own experiential world, and worse, they have no faith in the value of their personal experiences. This actually comes from a habit the profession developed a long time ago. The film industry in its current state does not encourage filmmakers to seek out their own innermost, authentic voices, because those voices are linked to reality. This is why, from the beginning, I've maintained a distance from the industry. I've seen innumerable domestic films, but not a single one resonates with my own experiences. That's why I figured I'd better make films myself.

In those days I knew very little about cinema, unlike my classmates who came from cities with better communications and access to information. Where I was from we didn't have the rich variety of DVDs and VCDs available today. In those days in Shanxi, even books on the subject of film were few; the major works in the history of cinema came to us third hand and by word of mouth. Later, when I too began to follow the Tuesday and Wednesday crowds to the film processing plant where we could watch "internal reference films,"[7] I couldn't help but find this unfair. Film resources had become a treat only a minority could enjoy. The

---

7. "Internal reference films" (*nei can pian*) were films, domestic and foreign, with restricted viewing mostly to cadres and film professionals. Jia explains and deplores this 1950s "invention" in another text in this book, "The Unstoppable Movement of Images."

monopoly of resources had transformed cinema into a privilege. It's as if the right of the masses to express themselves through cinematic images had been abolished, and those who had managed to elbow their way into the tight film community went on to uphold the mystique of cinema. I often wonder why cinema can't simply be like literature or painting: an option one might choose to express oneself. That's when I began to embrace the idea of independent cinema and to get an intimation of how I would orient my work in the future.

Sun: During your time at the Film Academy, you established a youth experimental film group. The core members of the group were all students from the literature department, and some even came from outside the Film Academy. In a film system that places film directors at its center, wasn't that some form of contestation? And what was its significance? The members of your team were then successful in a "revolutionary" way, but are they still involved in new film practices?

Jia: From the beginning, and still today, I've always thought that cinema is collaborative. Since the development of DV technology it's become much easier to get access to films, but that doesn't mean that filmmakers should be lone rangers undertaking the entire production from cinematography and sound to final editing. Each of these roles requires specialized knowledge and they need to be done by professionals. I don't deny that some people might be well versed in all of these areas, but the idea of so-called auteur cinema is incompatible with the very nature of film production. It also shows a fear of collaboration. The advent of DV production has generated a new myth. Two years ago, I wrote an essay entitled "The Age of Amateur Cinema Is about to Return." Those who read the piece know that I view amateurism as a force to counter stale creation and not a lowering of the standards of filmmaking. When a new film technology arises that shatters myths about filmmaking, in no time it's spun into a new myth in China. That's why I believe in defiance as a way to change things.

In 1995 when, along with my classmates Wang Hongwei and Gu Zheng, we established a youth experimental film group at the Beijing Film Academy, our main motivation was that we wanted to form a collaborative unit. We all wanted to engage directly in filmmaking, but we didn't know how to go about it. Already, at that time, we had abandoned any hope of working within the system, and we wanted to find some way to make films outside the film establishment. The members of our group came from different departments of the Film Academy: there were people from all branches, including cinematography, sound, and production. We were quite open and we accepted talented people from outside the Film Academy. This open attitude later proved to be quite beneficial to our development. When I eventually began to work transnationally, I realized that I had relatively little difficulty working with people from diverse cultural backgrounds.

With limited funds raised by the members themselves, we began our film practice. Some money came from remuneration I received for writing television dramas; some money, friends donated. Gu Zheng's mother in Shanghai even sent us some basic spending money. We collected what I considered quite a bit of money. We shot two films, *Xiao Shan Goes Home* and *Dudu*, which allowed us to gain experience in every part of the filmmaking process: conception, fundraising, equipment rental, on-location shooting, post-production, up to the actual screening of the film, promotion, and even giving public talks. To a degree, film is empirical work and the more experienced you are, the greater your ability to control the process. I have to say that, before we shot our first feature-length film, *Xiao Wu*, we had groomed ourselves to do it. I've often said that you must persist in finishing a work. You can't stop filming in midstream because you feel it's no good or abandon during post-production. Even if the film is really lousy, you have to hand it over so others can see it. You have to go through the entire process of filmmaking. I also believe that we should not despise short films. You have to take a deep breath and complete the work. Short films are really good training and they allow you just as well to manifest your talent.

Of the members of the group, Wang Hongwei, Gu Zheng, and myself have maintained the longest collaboration. Wang Hongwei has played the main role in *Xiao Wu* and *Platform*; he's very well liked, even Kurosawa's producer Teruyo Nogami is a fan. The last time I returned from Japan, that elderly lady asked me to deliver a present to him; she had also made a cartoon of *Platform* and inscribed it: "My beloved

Wang Hongwei." Gu Zheng, for his part, has pursued his studies up to a PhD; he has participated as a writer and assistant director on all of my films. The sound engineer for *Xiao Shan Goes Home* and *Xiao Wu*, Lin Xiaoling, is studying montage in France. One of the two producers, Zhao Peng, is now working at the Beijing Film Studio, while Zhang Tao is doing graduate studies at the Film Academy. Wang Bo, the artistic director of *Xiao Shan Goes Home* and *Dudu*, is absorbed in his own artistic practice; his recent Flash animated productions are quite masterly.

Sun: Following the success of your first work, *Xiao Shan Goes Home*, you and the members of your production unit started to become known outside of China. And after *Xiao Wu*, you were admitted into the global art scene; you started to make regular appearances in international film festivals. From a small county town in the hinterland to Beijing, and then onto the world stage… This kind of spatial leap is also, in current parlance, time travel. As you were crossing these borders, what kind of special awareness did you have? Do these three spaces, now that you have experienced them directly, seem more distinct or are they blurred?

Jia: In the case of *Xiao Wu*, which was shot in 1997, part of the investment came from Hong Kong and the rest from an advertising company in Shanxi. This can be considered a regional collaboration; the executive producers from both parties had only just begun investing in film. But when I made my second film, *Platform*, it was already a multinational investment venture; there were investors from Hong Kong, Japan, France, Italy, and other countries. This doesn't mean that the investment for *Platform* was huge or that the individual investing firms didn't have the capacity to be the sole investors; it was a transnational market necessity. This joint investment model is the basic structure for the markets of the future. It's fundamentally an optimal method: it reduces the investment risk, and it allows a director's work to be shown in more countries, to a wider and more diverse audience. Cinema is a declining industry. In the long run, artistic film creation simply cannot rely on one country's investment or one country's market to survive. It has to be said that, as a director, early access to the global art system helps you maintain continuity in your creative work.

What's even more important to me is that I'm no longer limited to a region. I have the opportunity to get acquainted with the work of my peers from other countries in a timely fashion. This helps me, as someone with relatively high standards, to ceaselessly criticize my own work. I've always maintained that a filmmaker who cares about the state of cinema should have an understanding of film history and of his contemporaries' work. No one's work can exist in isolation.

Moving from Fenyang to Beijing, and on from Beijing to the whole world, has made me realize that the lives of humans are very similar. Regardless of differences in culture, food, and traditions, people all face the same issues. We are all born, we get old, sick, and die. We all have a family. We all have to deal with the passage of time and bear the same ineluctable fate. The universality of the human condition makes me cherish even more my own experiences. I also believe that what my films offer isn't some oriental spectacle of a remote small town in Shanxi and my films are not about political repression or the state of society; they're about the crises involved in being human. I'm more and more convinced of this.

GRAY ZONES BETWEEN LIFE AND ART

Sun: For many years, there's been a tendency in the art world to endow artists with mythical status, to elevate them above day-to-day life, and to turn them into otherworldly creatures. For the sake of art, we imagine them capable of anything. For instance, we hear of Soviet Russian poets who, having no bread to eat, nevertheless refuse to abandon their poetry, or the artist who, unable to pursue his artistic practice, doesn't hesitate to commit suicide. Such talk may very well poison young people's minds.

Recently I read a collection of Van Gogh's correspondence. In the art world, Van Gogh is a classic example of someone who has been accorded mythical status. And yet, in every letter he wrote to his younger brother, Theo, even though there is always some discussion of Van Gogh's views on art, the dominant topic is money. Van Gogh goes on about how he has spent all his money and shyly asks Theo to send more cash, or he promises his younger brother that his paintings have every chance of bringing enormous profits. This picture is probably much closer to artists' basic situation than all the usual lofty discourses. As someone involved in film, your difficulties arise not only from the pressures of everyday life, but they must also involve the necessity to find funding for the production of your work, not to mention the social pressures that may or may not allow you to actually make the film. I imagine that, in the course of all of these pressures, there must be difficult moments when you have to "play the helpless poor relation," when you have to set your principles aside, to pretend, lie, and dupe others, and so on. It seems to me that such situations requiring compromise may have more educational value than the times when you don't bend. Could you, without getting yourself into trouble, talk about similar experiences you've gone through and how they affected you?

Jia: Compared to literature, painting, or music, a film is not an easy project to launch. For example, if a writer suddenly gets an impulse to create, he finds a quiet place, lays down some draft paper, and he can enjoy the pleasure of writing. But a film director, when he's inspired, must first contain his excitement and throw himself into a great mountain of tasks that have nothing to do with creation: finding money, meeting with all sorts of people, explaining his script countless times. By the time you stand behind a camera ready to shout "Ready… Action!" one or two years may have gone by. After six or seven hundred days of talking about that moment of excitement you had a few years back, can you keep the faith in your inspiration?

What's most problematic is that often during this endless waiting period, unfortunate encounters with people or situations will intervene in your life. Will these people and situations affect your conduct? Will they shake your values so that suddenly you move away from your original creative intention?

When I was at the Film Academy, I often acted as a "sharpshooter;" I'd sit in the study room and quickly scribble television dramas for others. In those days, I was happy to do this work, because it allowed me to earn a bit of money to subsidize my own short films. One day a friend I had known for a long time came looking for me; he wanted me to help him write a twenty-episode television drama. He was a Beijinger and claimed that members of his family were the investors, so money was not a problem. During the writing period, he was so considerate that it moved me. But once I'd written the last word and delivered the final episode of the drama, he suddenly became hostile and said he wouldn't pay me. He also insulted me, said I was a dumb ass for not having signed a contract with him. Some time before this, he had left two boxes of cups he couldn't sell in my lodging. Now he told me, "I don't want these cups, keep them in lieu of payment." To this day, those cups remain on my balcony; I often stare at my porcelain remuneration, which has made me profoundly distrustful of people.

Much later, in order to obtain my filming permit so that we could begin shooting *Platform*, I was obliged to return several times to the Film Bureau. One day I finally agreed to pay a fine of RMB ¥10,000 [USD $1600] relating to *Xiao Wu* and wrote a clear, straightforward self-criticism acknowledging that I had indeed disrupted the normal flow of my country's foreign cultural exchanges. When I left the Film Bureau, I suddenly thought of a line by Bei Dao: "I do not believe!"[8] In my mind, the affair of the cups and my self-criticism were linked. As the bus crossed Chang'an Avenue, my heart had sunk as low as it could go. At that moment, I didn't even believe in myself.

Sun: Another myth about artists is that inspiration is forever gushing forth, as though their heads were volcanoes. In reality, for all people who have the experience of artistic creation, once the initial inspiration has passed, creation is a long process replete with dry spells that you have to overcome. Inspiration and talent are admittedly important, but in the end, the decisive factors may well be persistence, hard work, willpower, and your ability to relieve the pressures of creation. Having said this, I'd like to invite you to describe your state of mind when inspiration dries up and tell me about your methods for dealing with that and the general pressures accompanying creation.

Jia: I think none of us is like those talented silent movie directors who, during cinema's infancy, really carried out truly creative work in film. Every creation may be new, but figures like F. W. Murnau basically invented

---

8. This defiant line is from the poem "Answer" (*Huida*), dated 1976.

all the ways of making cinema. I've always thought that from the 1980s on, cinema has been in decline everywhere, that the quality of films is constantly falling. Even Hollywood has lost the creativity of benchmark films like *Bonnie and Clyde* and *The Godfather*. All of cinema is in a dry spell.

I often feel frustrated with my inability to invent a new film form. This doesn't mean that I'm unable to come up with new visual approaches or that I've lost the ability to represent the passage of time. These together are the basic components of actual filmmaking. But beyond that, there's a film language, architectonics that constitute the medium of film itself. It's this language that we need to radically reinvent, like Sergei Eisenstein did. On rereading Eisenstein's writings, I realized that for the past few decades we have been rethinking film form—even Jean-Luc Godard's and Andrei Tarkovsky's revolutionary directions aren't radical enough. That's why cinema is at a standstill and we're still holding the old bullhorn in our hand.

Lately I've been looking mainly at earlier theory and old films for solutions to practical problems. Increasingly, I don't trust my former understanding of texts and films. I feel that somehow we've overlooked many of the principles discovered by great masters like Eisenstein or even Leonardo da Vinci. This is because we can no longer be like people of those eras, who were versed in all branches of learning, who were immersed in science while undertaking a kind of religious quest for new forms. This inability has damaged the quality of our cinema.

For me, creation is a process of continuous self-struggle. If my work proceeds too smoothly, I frequently become suspicious and restless. Sometimes, when I'm truly stuck, I'll skim through interviews of my peers. Then I realize that all creative work actually emanates from extreme perplexity.

Sun: Of all the arts, cinema relies the most on funding and the market. For filmmakers in China, resources for both are presently in relatively short supply. Every year fewer people wish to invest in film, and the resources of Chinese moviegoers are limited. Every year the opportunities to participate in international film festivals are limited, and so when one person gets an opportunity it implies that many other people have lost their chance. As far as I can tell, the film community in China is fraught with tense and difficult interpersonal relationships; it's easy for people in that milieu to become mentally unstable. Under such circumstances, how do you maintain your own peace of mind to pursue your work without too many distracting thoughts?

Jia: My method is simply not to enter into that world; I'm not at all interested in their personal enmities. In Beijing, I can say I'm more or less a lone operator. Although I'm somewhat isolated, I can focus and absorb myself in my own work. From the very beginning, I've had a more or less comprehensive plan for my work: I hope to gradually build a world of my own in film. This is a compelling way to work, which makes me pay very little attention to things outside my creative work, including wins and losses at film festivals and good or bad ticket sales, because neither are my ultimate goal. What I am always concerned and anxious about are artistic questions, and those are mine alone. They have nothing to do with the film community or anybody else.

There's so much at stake in the film industry, which is why there is competitiveness between peers. Survival of the fittest is the very cruel name of the game. The day I lose my creativity, I will leave cinema. From that point of view, cinema will not be my lifelong occupation. Making a film gives you a kind of primitive joy: the joy of a creative process that is fabricated out of thin air. I think of that first moment the Lumière brothers saw the moving images they had filmed. It's that joyful instant that our little works offer us.

That everyone might share that same joy? That would simply be asking too much.

Sun: Film requires the involvement of many people, and they must work in a concerted manner. It's not possible that everyone will share the same opinion on any given matter. Could you talk about some disagreements that you and your team have encountered during the creative process and how they were resolved?

Jia: Most of the people with whom I collaborate have been working together from the onset, since *Xiao Wu*, and some even since *Xiao Shan Goes Home*. We're quite a stable team, but we're not at all closed-off. In the past few years a whole slew of new friends has joined us, which has helped to raise the quality of the films in every aspect. As the film director, I get too much of the credit; their skills and contributions to the films are rarely recognized. This is unfair to them and it often makes me feel guilty.

Take the case of Wang Hongwei. He has played the main role in every film, from *Xiao Shan Goes Home*, *Xiao Wu*, to *Platform*. He has quite a big following in Europe and Japan, but at home, because our films were banned, not many people have had the opportunity to admire his performances. Since he has very little exposure in the media, other Chinese directors have no way of assessing his box office appeal, so he is rarely invited to act in their films. Another case is my cinematographer, Yu Lik-wai who, already after *Xiao Wu*, was in demand by several transnational productions. In the space of a single year, he was invited to work on three South Korean films. But at home, because very few people have had the opportunity to view our films on the big screen and have only seen them on pirated VCDs, they wrongly believe that *Xiao Wu* was shot with a home video recorder.

We have all known and worked with each other for more than five years; the closeness that comes from such a long acquaintance tends to minimize those clashes that are hard to avoid when working together. Aesthetically, we are amazingly in agreement; often our excitement over something becomes contagious. Most of the time, we are all enjoying the pleasure of working quietly on something we've agreed upon. On

IMAGES TAKEN FROM THE WORLD OF EXPERIENCE

the set we interact very little; what we need to say has all been said around a tofu and fish hotpot at the outdoor restaurant by Beihang University. In the making of my films, there are very rarely any divergent opinions on the direction; they all understand my inner world and identify with my set of values.

The only time there was a conflict was while recording sound in the post-production of *Xiao Wu*. My sound engineer at the time was Lin Xiaoling. She was a friend with whom I had collaborated on *Xiao Shan Goes Home*, and she was responsible for a large part of the production work during the period of the experimental film group. We also made the highly improvised film *Dudu* together; on the set she helped me finish the script and played the main role. I really liked her resourcefulness on the set. She was the one who found the music for the end of *Xiao Wu*. It was a hymn, sung in the Fenyang accent, that she heard by chance in the Fenyang church. But as we were mixing sound for *Xiao Wu*, I asked her to add in a lot of street noise. I kept saying, "Make it rougher… Make it rougher." She disagreed with me because, from a technical point of view, this violates the rules. She was very self-conscious and wanted to preserve her own sense of professionalism, but by then I was already stubbornly entrenched in my position. When I think back today, I realize I was despotic, never considering other people's feelings. When I asked her to mix in a lot of pop music, our disagreement became antagonistic. Lin Xiaoling left the recording studio without saying a word. I was at my wits' end. I finally sought out a classmate, Zhang Yangqian, to replace her at the last minute.

I think that our divergence stemmed from our different backgrounds. Lin Xiaoling's parents were both high-cadre intellectuals; as a child she studied classical music and had a good family upbringing. In her experience, the world is not at all so rough. As for me, a bumpkin who's had ups and downs, life is not that exquisite. Neither of us was wrong, but my stubborn insistence hurt her. I've never apologized. Today I want to say "I'm sorry."

I still think that a film director must encourage the free airing of views, but in the final analysis, cinema is an art in which the director is a despot with the final cut.

FILM CONCEPTION AND FILM PRACTICE

Sun: Since the 1990s, China's film world seems to have become obsessed with cinematic realism. Concepts and techniques such as the long take, on-the-spot recording, non-professional actors, anti-montage, and anti-dramatization are constantly being emphasized in university classrooms. Even a director like Zhang Yimou, who has a strong expressionist style, filmed *Not One Less* (*Yige dou bu neng shao*) in that vein. But from another perspective, even the most realist cinema cannot avoid rhetorical devices; even their longest long take is but a fleeting moment in the film's narrative. You choose those segments of time and not others; you choose high or low shots; you focus on filming your subjects with a frontal angle or in profile. All of these produce different meanings for viewers. Does this mean that realism in cinema is not realistic but is rather a way to appear more real than reality? As a filmmaker who quite likes realist techniques, in terms of truth in art, which is most important to you: to create an impression of reality or to technologically construct that reality? And how do you tell them apart? Which leads to another question: When you're making a film, are you more mindful of staying true to yourself or of convincingly portraying that reality to the viewers?

Jia: The Polish director Krzysztof Kieslowski once said something that touched me deeply. After having filmed a great number of documentaries, he said: "In my opinion, the closer the camera is to actual things, the more likely it's getting closer to artificiality."

It's quite possible that the so-called reality that is produced with realist techniques is actually obstructing and concealing the way reality really works. And the use of dialect, non-professional actors, natural settings, synchronous sound, and even long takes does not in any way represent reality itself. It's totally conceivable that people might use these same techniques as a recipe for a hallucinogenic drug that would transport you into a nonsensical world.

In fact, the real/truth in cinema does not reside in any particular moment; it only lies in the overall design. That's what motivates the beginning, continuations, transitions, and ending. The tight inner logic yields a convincing order of the real that remains after the narrative structure is deconstructed.

For me, all realist techniques serve to convey my perceived experiences of the real world. We almost have no way of approaching reality in itself, and in any case, the purpose of cinema is not simply to reach some standard of truth. I pursue a sense of the real in cinema more than I pursue reality itself, because I think that the sense of the real is aesthetic, whereas reality itself is the domain of sociology and sciences. Likewise, in my films, social issues are experienced as individual existential crises. After all, I am a filmmaker and not a sociologist.

Sun: I remember that when I viewed the three-and-a-half-hour-long version of your new film *Platform*, my impression was that your creative mindset was not as relaxed as it was for *Xiao Wu*. There was a tendency to overstate things. Is it because the idea of "waiting for something to happen" is for you a very serious topic? You also seemed to be very much in contact, before and during the shooting, with critics and intellectuals: Did they cramp your style to some extent? How do you look upon your relationship with the world of critics?

Jia: I had the script for *Platform* before filming *Xiao Wu*. It was 1996 and I was still in school. There was a depressed and inhibited atmosphere in Chinese intellectual circles at the beginning of the nineties, and a rapid advancement of the market economy that despised and debilitated culture. It reminded me of the eighties. My own youth, from the age of ten to twenty years old, took place during that period, which was also a decade of dramatic change in Chinese society. My troubled youth was intertwined with the entire country's fast-paced development. I felt compelled to recount my own experiences against the background of that era. If one can say that film is a way to recollect, then what we portray on our silver screens are almost all official recollections. And there are those who ignore mundane life, despising everyday experiences to concoct historical fantasies. I stay away from those two types of film. I wanted to relate the feelings that are entombed in the past, those impulses that have remained unfulfilled, and those personal experiences that have nowhere to go.

Platform was overwhelming, but I knew that it was going to be a rather huge undertaking. When I got my first opportunity to make a feature-length film, because financial resources were limited, I had no choice but to set *Platform* aside for a while and first draw a sketch of a "Xiao Wu" youth from a county town. From start to finish, I find that *Platform* and *Xiao Wu* constitute a diptych. *Xiao Wu* is the sequel to *Platform*, and *Platform* is the source of *Xiao Wu*.

In the winter of 1999, we started to film *Platform* in Fenyang. As some critics have noted, this work is both restrained and untrammeled. For me, it was also a very special creative experience: during the entire shooting I was inwardly torn. From start to finish, I struggled to find a balance between myself and the past, between myself and the film. Using the camera to pursue memories and improvising on the spot often clashed with the precise planning required to make a period piece. During shooting, I know I seemed terribly headstrong, but the workers tolerated all this. They understood how difficult it was to capture that subtle, gossamer-like atmosphere concealed within slow passages of time. I needed time to recapture those flashes of memory that had been so crucial.

I know that, in *Platform*, I've overstated things at times. That's because I have more memories than I can manage. The film is a rock pressing down on my heart: once I've removed it, I'll be able to start a new project.

As you can see in the long list of acknowledgments at the end of *Platform*, I have many friends in intellectual circles and among critics. Poet Xi Chuan played the role of the leader of the cultural troupe in its early days as a work-unit, and painter Song Yongping played the boss when the troupe begins to work for profit. They both have more artistic experience than I do. They have maintained their creative drive for a long time. Sometimes I find myself in despair; I lose self-confidence and all interest in cinema. This faint-heartedness that is hard to share, these negative moments that come back again and again weaken my will to go on. That's when I think of others and don't feel so alone.

I think any creator seeks critical responses. Choosing cinema as your medium means that you have chosen a form with broad exposure. But in the final analysis, I don't think that criticism can fundamentally influence me, because criticism cannot guide my work: I'm at the helm.

Sun: I hear that you've recently made a documentary. Can you tell us something about it?

Jia: The documentary I've just now completed is called *In Public*. It was made as part of a film project with the Taiwanese director Tsai Ming-liang and British director John Akomfrah. Each one of us made a thirty-minute digital video on the theme of "space." It's called "Digital Short Films by Three Filmmakers" and was funded by South Korea's Jeonju International Film Festival.

My own part was shot in Datong with very little dialogue and with some distance from people. It's an impression of a city. More and more, I find that digital technology is better at conveying the conceptual nature of documentaries. When you enter a space, nobody pays attention to your presence. Everything moves forward naturally; you can capture the natural order of things, and that's how abstraction is produced.

Sun: And do you have a new film project in the works? Can you tell us a thing or two about it?

Jia: I'd like to film the story of today's youth, also set in Shanxi.

Originally published in *Avant-garde Today (Jinri xianfeng)*, volume 12, 2002.

THE GENETIC COMPOSITION OF MY FILMS   17

Translated by Alice Shih

When I was seven or eight years old, my father often told us about his experience on a film set. The Cultural Revolution had just ended, and there were a lot of meetings about reconstruction plans. Smiles started to reappear on people's faces after a long period of oppression. It didn't matter how late my father returned from school, the whole family would gather at the dinner table to eat and chat. Sometimes the power went out, but we wouldn't light candles. The fire of the stove lit our faces as we immersed ourselves in the warm atmosphere in darkness. My father would patiently describe the details of filmmaking to us on such occasions. When he was a student at Fenyang Middle School, he heard that the Changchun Film Studio was making a film just outside the city near a village by the valley, so he went with his classmates through hills and creeks to take a look. They stood near the camera crew and watched them scrambling with the tripod to find suitable spots for the camera. Some people dressed in peasant clothes were chiseling and building ditches in front of the lens. My father and his friends thought they were at the wrong place at first, taking the commotion for a team of geologists at work. Then they discovered that these peasant workers were following the orders of one person. He realized that the commander must be the director, and they were indeed making a film. So this group of youths rejoiced and stood quietly in the valley and watched filmmaking until sundown. They reluctantly left as the crew wrapped up at the end of the day, but he seemed to have grasped the secret of filmmaking. The fiery stove cast light and shadows on his face as he said, "You need light to shoot a film."

It appears to me that my hometown of Fenyang in the province of Shanxi is blessed with the special gift of light. It is situated on a high plain of yellow earth. Every afternoon, an abundance of direct sunlight wraps the whole town and countryside in warm vibrant colors. This picturesque landscape makes people poetic and romantic. The seasons here are distinct and very suitable for filmmaking. The film that my father saw in the making is *Our Youngsters* (*Wo men cun li de nian qing ren*) (1959), written by Ma Feng, a Shanxi native, and directed by Su Li. The story is about a group of young people in Shanxi who want to divert water through the mountains to build a power plant. Since Fenyang is considered one of the two hometowns of "Potato School"[1] author Ma Feng—as he was placed here to experience peasant life after the liberation—the director had decided to shoot on location here. What they didn't know was that while they were scrambling to make the film, they had piqued the interest of the young bystanders. When my father told me repeatedly about his film set experience, he did not know that he was passing on a filmmaking gene to his son. The influence of cultural inheritance is hard to comprehend. When I shot my first feature, *Xiao Wu*, in Fenyang in 1993, I suddenly realized that my career choice to become a director was definitely related to my father's experience.

Originally published in *Chengdu Economic Daily* (*Chengdu shangbao*), August 10, 2006.

---

1. The Potato School is a contemporary genre of literature founded in the fifties to mid-sixties by author Zhao Shuli. Authors of this school were born and raised in peasant families in Shanxi Province. Their works embrace a rich reflection of country lives.

MAKE FILMS ACCORDING TO YOUR OWN BELIEFS   18

A Conversation between Hou Hsiao-hsien[1] and Jia Zhangke
Translated by Alice Shih

Hou Hsiao-hsien: *Still Life* and your documentary *Dong* are they not similar?

Jia Zhangke: Totally different. I started with *Dong*, and after shooting for ten days I suddenly had the urge to shoot a narrative film.

Hou: Was that because of the inspiration of the location?

Jia: Right. During the course of the documentary shoot, I kept imagining creative narratives in my sleep. The location, the space, and local faces are very different from those we encounter in northern China. Their daily struggles are also different. In Beijing or Shanxi, you might be poor, but you would always have some sort of small electric appliances, a few storage boxes, closets, and pieces of furniture. In Sanxia, on the

---

1. Hou Hsiao-hsien, one of the most important filmmakers of the Taiwanese New Wave Cinema, began directing in 1980. In 1989, his film *A City of Sadness* (*Bei qing cheng shi*) won the Golden Lion at the 40th Venice International Film Festival. In 1993, *The Puppetmaster* (*Xi meng ren sheng*) won the Jury Prize at the Cannes Film Festival. Through his broad and warm visions of life, the earth, history, and memory, he blends the elements of sight and sound in a glorious way.

other hand, many families really have nothing in their homes, barely four walls, if you are lucky.

Hou: My creative process is similar to yours: first the encounter, then come the thoughts. I was watching *Xiao Wu*, and I saw, by your execution with the actors and the camera, that your experience gives you insights. After *Xiao Wu*, you were recognized, and you seemed too rushed to pour out all the brewing thoughts from your head into your follow-up works. And it seemed like you put people aside, emphasizing space more. To me, that seemed too forced and apparent. With *Still Life*, I see a return to real people in everyday lives. It's a direct reaction to reality, and it soars with the same kind of energy you had in *Xiao Wu*.

Jia: I've made three films between *Xiao Wu* and *Still Life*; I was having some baggage indeed.

In *Xiao Wu*, I was concerned with the physiological impact of human emotion. The subsequent films were basically a reflection on the human condition through history—how we establish our position in a relationship, and how we have all lost a bit of our sheen.

In Sanxia, the immediate reaction of the scorching sun on our bodies brought back things I had lost. I was especially affected when I visited the demolition sites. I saw workers moving bricks with their bare hands, brick by brick, until the whole city disappeared completely. The people inside the frame moved me, and the wild blood inside my body boiled once again after years of civilized living in the city. It was plugging a hole my creativity had been leaking out of. My creativity became wild and active again.

Hou: It's evident that one cannot create out of the imagination alone; we need reality.

My situation was a bit different from yours. After I finished *Flowers of Shanghai* (*Hai shang hua*) (1998), I was waiting for a pitch. I wasn't sure what I wanted to shoot next, and I didn't mind not knowing. I know that I have the skill and I have also accumulated a lot of ideas. I was ready to take on challenges from others. That kind of creative challenge can be very interesting as well.

THE MARKS OF LIFE ARE POWERFUL ONCE WITNESSED

Jia: When I was still learning filmmaking, I found *The Boys from Fengkuei* (*Feng gui lai de ren*) (1983) very enlightening. I remember seeing it back in 1995 while I was at the Film Academy, and I was stunned. It hit very close to my heart, like it was shot with my pals back home, yet it was about youths in Taiwan.

I realized afterward that life always leaves a mark on people through experience, and its latent power is liberated when this experience is captured through a story and witnessed. In my generation, which was born during the Cultural Revolution, art was basically a calculated mix of pop and legend, the two essential elements of revolutionary arts. It's popular so that the masses will understand the political message and legendary to eliminate reality and self. What's left is only a fable, like the story of *The White-haired Girl* (*Bai mao nu*) (1950), which depicts a woman who has lived in a temple for thirty years, until all her hair turned white, and is finally rescued by the communists… The story has no portrayal of daily living, not even domestic life; it's totally detached from life experiences.

When I saw *The Boys from Fengkuei*, I felt endearment and familiarity. It was the same with *A City of Sadness*, a Taiwanese story about the 2/28 Incident,[2] which I knew nothing about. I had no difficulties taking it in; it was just like interpreting calligraphy to me. I have learned and inherited your techniques and your narrative language.

Hou: Creativity is tied to your early encounters. It comes from the root. For example, I was greatly influenced by literature. I was reading books by Chen Yingzhen,[3] like *A Tribe of Generals* (*Jiangjun zu*), *Bell Flowers* (*Lingdang hua*), and *Mountain Path* (*Shanlu*). These books are all about the era of White Terror, when the Kuomintang oppressed the people. They helped me develop my views and attitudes about history.

That was one period I went through. Afterward, I got back into human interests. After *Flowers of*

---

2. The 2/28 Incident, a violent result of an anti-Kuomintang government uprising in Taiwan on February 27, 1947, resulted in a massacre of close to thirty thousand civilians. It commenced on February 28, or 2/28, followed by the start of the Kuomintang's White Terror period in Taiwan, when thousands more people on the island were killed or imprisoned.

3. Chen Yingzhen, b. 1936, is a renowned Taiwanese author who was a political prisoner from 1968 to 1975 and is still actively involved in the Chinese literary circle today.

Shanghai, I came back to contemporary times, and my subsequent films, *Millennium Mambo* (*Qian xi man po*) (2001), *Café Lumière* (*Kôhî jikô*) (2003), and the most recent one, *Flight of the Red Balloon* (*Le voyage du ballon rouge*) (2007), are all on contemporary issues.

ADJUSTING THE GENRE TRADITION AND EMOTIONAL EXPRESSION

Hou: Recently, I realized that filmmaking involves both desire and reality. World cinema leans toward a dramatic reality. Chinese filmmakers' attention is not entirely on the narrative form but more on the form of emotional expression and artistic conception. Therefore, our aesthetics may not be acceptable to the wider public.

Jia: It is the same in Mainland China. The Chinese tradition of movie watching is about drama. A movie is synonymous with drama. The general public goes to the theater with a high expectation of dramatic twists. They may not care about the quality of the film, as long as they have a plot to follow. Alternative films without strong plots find it hard to fight this audience mentality.

Hou: Western films sprang from the dramatic stage. This is such a strong tradition. Film used to be a visual medium open to various forms of expression, but after the introduction of sound, film has gone back to serve drama. Playwrights were brought in to write film scripts and to emphasize the plot. Because of the domination of these practices, I have to defend my own way of narration. My form of expression is more like the ancient text of *Classic of Poetry* (*Shijing*); its intent is not storytelling. But people find it hard to understand, as they have been profoundly influenced by Western theater.

This trend seems impossible to break, but if you understand this alternative film form, you can find a workable space to adjust the ratio of traditional genre and emotional expression, to create a harmonious mix of Eastern and Western culture.

REMOVE THE UNNECESSARY STUMBLING BLOCKS

Jia: I find that film has also been affected by new visual technologies, such as DVDs, electronic games, and satellite television. Watching Taiwanese TV, I discovered a wide variety of programming—some on crime and others on political controversies. The whole society is already so dramatic, how can we make films even more enticing? It seems like they don't bring much to the table. But I'm glad to see some directors who are able to find a way to express their views through genre films, making use of the genre package. After all, genre elements are well received.

Hou: Good genre films are also inspired by reality; it is the common denominator of all good films.

Jia: I remember last time we chatted in Beijing; you said something that is still on my mind. It was, "Use the simplest method to say the most." My interpretation of the simplest method would be to do away with any unnecessary demands that would alienate the audience.

Hou: Right, be direct. If you focus on the narrative, the story will flow naturally and it will be easy to grasp. In this way you can reflect on your abilities to observe, assimilate, and react. I do find that it is really challenging to portray deep issues in a simple way—simple enough to be accessible to the public but encompassing deep meanings at the same time. This is not so easy.

Jia: "Simple" could mean a very direct approach, like the neo-realist Italian films made in the late 1940s, which audiences enjoyed. *Bicycle Thieves* (*Ladri di biciclette*) (1948) proves that audience approval does not preclude deep meanings. Fellini's *La Strada* (1954) was also well received.

On the whole, our thinking about form versus content is totally fogged up. We need to recalibrate and find a direct and simple approach.

RETURNING TO THE ORIGINAL UNCORRUPTED MINDSET

Jia: What's your view on the work of the new Taiwanese directors?

Hou: They have watched too many movies growing up. They have fallen into the trap set by films they saw ever since they made their first shot. They are shooting "films within the world of films." They do not know much about real life experiences, real affection, or their positions in life. But not only that; they are just not bold enough, and their form and content are easily swayed by traditional images. If they were valiant, their films would show more conviction.

Jia: That seems to be a very common problem.

When I first started making films, the Fifth-Generation directors were undergoing some changes. There were a lot of discussions in China at the time trashing the cultural values of film. They emphasized the importance of filmmaking as an industry—the investment should entail reasonable financial return. I felt sad, because after a film got released, people did not discuss the message conveyed by the work but rather its economic value.

Therefore, I believe in the importance of the "individuality" of a director. You need to be robust to fend off outside influences. Early filmmaking was like juggling, and it should be fun. It should not be about extrinsic factors. Filmmaking should return to the original straightforward mindset.

I know it's easier said than done. Like you were saying earlier, it took me a while from *Xiao Wu* to *Still Life* before I got back this lost feeling.

Originally published in *Elite Reader (Chengpin haodu)*, December, 2006.

IT'S A NARRATIVE AS WELL AS A DOCUMENTARY   19

A Conversation between Tsai Ming-liang[1] and Jia Zhangke
Edited by Tan Zhengjie and Zou Sin Ning
Translated by Alice Shih

DIRECTORS WHO HAVE FALLEN THROUGH THE CRACKS

Tsai Ming-liang: I think that our concepts and attitudes about film are similar. We don't do commercial films, and we seem to be different from the mainstream market mentality.

The flavor of your work reminds me of the film *West of the Tracks* by Wang Bing, which I saw at the International Documentary Film Festival of Marseille. They deal with the rapid changes taking place in China, which a whole generation of young Chinese directors is facing. As a result of these economic and socio-political changes, most Chinese people are chasing money. Ideals and dreams are disappearing. Your generation of directors sees the tension. In my eyes, your generation has unfortunately been overshadowed

---

1. Tsai Ming-liang is a Malaysian Chinese filmmaker who works in Taiwan. In 1994, his film *Vive L'Amour* earned the Golden Lion at the Venice International Film Festival. Tsai has established himself as a fresh and unique voice in the world of Chinese films. His films mostly focus on isolation, loneliness, human separation, and barriers.

by Fifth Generation directors such as Zhang Yimou.

Our situation is similar. My position in the chronology of Taiwanese filmmakers is after director Hou Hsiao-hsien and before the current generation of new directors. I am also a foreigner. We both have fallen through the cracks. We have our ideals but we struggle against reality in a big way.

Jia Zhangke: I agree that we have fallen through the cracks. Our work, our ideals, and our spirituality are totally different from consumerism. I was able to make my first film in 1997. That was the year when the Chinese consumerist economy took flight. My passions and ideas about which direction films should go were heading in the opposite direction from society. Unintentionally, I became a rebel against commercial culture and consumerism.

Filmmaking was not my first choice. I was born in 1970 and the Cultural Revolution had just ended when I started grade school. Material wealth was nonexistent at that time. Half of my childhood memories are about hunger. This extreme state of starvation is totally incomprehensible to the generations born after the eighties and nineties. They never had such an experience.

We did not have many options. For us kids, films were something we could enjoy once in a while, but it never crossed my mind that I would one day become a director.

We were living under tremendous pressure at the time. I felt something was trying to emerge from me, like a sense of destiny. We graduated from elementary

IT'S A NARRATIVE AS WELL AS A DOCUMENTARY

school after grade five, and things changed a lot. I continued my studies and started junior high. Some physically more developed classmates who had family connections quit school to become soldiers or policemen. Other classmates quit because their parents thought that a fifth-grade education was good enough and they entered the workforce in all kinds of fields. There were others who left school and were unable to find work, and they gradually became hooligans. Even though I was so young, the changes affected me. I felt that people are very different. Witnessing so many different journeys, I began to develop an interest in different kinds of people.

Because of this instinct, I started reading a lot. One year, I read Lu Yao's *Life* (*Ren sheng*).[2] It's about a very important social issue: residency registration. Basically, all Chinese citizens are divided into two categories—city or rural residents—and there is no mobility between the two. However, there is a bridge, and that is the high school graduation exam. I was only a kid then and wasn't aware of the injustice. After reading the book, I suddenly realized why city kids like me were fooling around all day, but our classmates who came from the villages were eating tasteless plant roots to stay awake studying until midnight. They were hoping to change their destiny!

I'm very thankful for reading. It helped me think deeply about things and to become skeptical. This is the reason why my films are all about the way that

---

2. Lu Yao (1949–1992), born Wang Weiguo in Shaanxi, is a Chinese writer who grew up in poverty. Lu Yao, his pen name, means "Far Road." He published his novella *Life* in 1982 and it was made into a film in 1984 by Wu Tianming.

individuals struggle within society.

Tsai: Other elements of our films seem to be the results of our life experiences—for example, when I feature Bai Guang and Grace Chang[3] songs and you use Cantonese and Mandarin pop music soundtracks. These elements are so important that the films cannot do without them. I don't think market-oriented directors would do the same thing.

I ran into two young Chinese students the other day. They saw films of mine like *Vive L'Amour* and *Rebels of the Neon God* in the nineties and wondered why the pace was so slow. Lately, they watched them again and said the slow pace didn't bother them anymore. They thought it was refreshing since everything is going so fast these days. It was the Taiwanese New Wave film movement of that time that allowed me to have such a creative space. There was no creative pressure and I didn't have to serve the box office.

Watching your films, I think you have achieved a rare and noble feat. Your thoughts stay on course and you don't bow to the market. You pursue your own interests when it comes to filmmaking, and in that way we are very much alike.

Of course, people will wonder what the heck we are doing. But it shouldn't be so hard to understand. Since there are so few people doing this type of work, it should be valued. We should really question the things that the majority is doing.

---

3. Bai Guang and Grace Chang are famous Chinese singers/actresses from the late 1940s to early 1960s with a long list of Mandarin pop chart toppers.

IT'S A NARRATIVE AS WELL AS A DOCUMENTARY

Jia: I couldn't agree with you more. Cultural work should be diverse. On this note, in the late 1990s when the economic boom started in China, values were homogenized. The public has changed and based its value judgments solely on money.

This is very bad for Chinese culture. As the saying goes, it's "demolishing peaks to fill valleys." I think it's the same in Taiwan. I've heard some critics say that directors like Hou Hsiao-hsien and Tsai Ming-liang have lowered the bar for Taiwanese cinema. I've also heard some say that Jia Zhangke has misled the youth and smothered the development of Chinese cinema. These are odd things to say. Peaks are peaks. Valleys can be left to those who excel there. Directors should specialize in their own genres. I don't believe that we have a duty to the film industry. Everybody should have the independence to do what he likes.

DOCUMENTARY AND THE OPTIMAL VIEWING DISTANCE

Tsai: You make narrative films and also documentaries. I've been making narratives all along and didn't give documentaries much thought. Opportunity knocked and I got the chance to make my first one, about AIDS, and then *A Conversation with God*, about shamans.

My film about AIDS was more of a traditional documentary. AIDS was still widely discriminated against at that time, as people thought it only happened among gay men. When the Fubon Charity Foundation approached me, Sylvia Chang, and four other directors for this project, I chose to do my segment on gay people. I wanted everyone to know the reality and to smash the stereotype. When the Jeonju International Film Festival invited me, you, and British director John Akomfrah for a series of shorts, I agreed and made *A Conversation with God*. At the beginning, I wasn't sure how and what I would be shooting with this little digital camera, and I was a little resistant. I wanted to shoot what I'm familiar with, so I chose Taipei.

It didn't feel like a documentary; it felt more like an emotional portrait or a diary. I shot what the environment made me feel at the time. The news said that fish were dying in the river, so I went to shoot the fish. I was curious about shamans and the function of underpasses, so I explored them with my camera. I turn visions into images, from which symbols arise. An underpass is a route to another destination; what could be the connecting passage between humans and God?

Overall, I haven't seriously considered making a documentary. I mostly make narratives and I don't like people saying that my films are like documentaries.

Jia: I made my first documentary in 2001 when I accepted the Jeonju International Film Festival's invitation to Digital Short Films by Three Filmmakers. On that occasion I made *In Public*. I was in Seoul and thought of shooting a concept film about space. I remembered when I was traveling and saw all these small- to midsize city facilities, for example some old long-haul bus stop. You can see how it functioned before, but now it

IT'S A NARRATIVE AS WELL AS A DOCUMENTARY

has been turned into a dance hall. Some middle-aged or jobless people might go for a dance after breakfast and then go have lunch. They might go back again after their afternoon nap. I was attracted to this transformation of spaces, but I didn't give it much thought or really observe it thoroughly. I just brought the camera to these public places and tried to perceive as much as possible.

I shot *Dong* after *In Public* and then *Useless*. *Dong* is about a painter that I'm very fond of named Liu Xiaodong. I was not familiar with how he works and I really wanted to know more, so I chose to make a documentary about him.

Useless is about Ma Ke and her fashion designs. I happened to bump into her product line and I was attracted to the philosophy of her brand, Useless. She criticizes contemporary China through her fashion. I never thought that fashion could make such an impact. Everything Ma Ke produces is handmade. She believes that in the past, people always knew where a product came from, like a scarf made by your mother or your sister. Handicrafts had sentimental value. After the introduction of assembly lines and mass manufacturing, this way of emotional communication was ruined. And then there were countless environmental issues. Ma Ke has a very comprehensive and reasonable ideology. She has the occasional paradox, but I value that.

I filmed Ma Ke right after Liu Xiaodong, one after the other. I was hungry to interact with contemporary artists. I was exploring how different media were facing the same reality in contemporary China. I felt that through this interaction, I could break from the constraints of film and enrich the cultural scene. I hoped I could use documentaries to help audiences appreciate and understand the work and thought of these intellectuals. It would be terrible if society silenced the voices of these intellectuals and artists.

Of course, I'm not saying that I accomplished this with just two documentaries, but I hope that we are ready to rebuild the link between intellectuals and the public. If we only sold a few hundred thousand DVD copies of *Dong* after *Still Life* in China, then probably only about one hundred thousand viewers really watched the film. But I thought it was important for these people to hear Liu Xiaodong's view on how the revolution, life, and our bodies are being treated by contemporary society.

Tsai: This reminds me of when the Louvre and the National Palace Museum asked me to make films for them. My background is not in fine arts, I thought, so why did they approach me? I figured it must be because my films are very observant. I have a sense of how artifacts should be observed, and I've always encouraged audiences to develop an observant attitude. This is exactly what a museum visit is about and is probably why I was invited to make films in these museums. Maybe they thought that I could draw film audiences to museums, and museum patrons might be interested in going to the cinema in return.

Speaking of the audience, when I think of a filmmaker behind the camera in front of his subject, I always try to find the optimal distance to fit my

IT'S A NARRATIVE AS WELL AS A DOCUMENTARY

reading of the situation. My films are not strict narratives; they aim to stir up thoughts and emotions. My audience is asked to observe and analyze, not to be too absorbed in the plot, as they would with a traditional narrative film.

My shots are limited; sometimes so limited that I have problems in the editing room. I just shoot what I want, and I keep all my shots. I think of all my shots before the shoot, but I have to make the most important decisions on the spot during the shoot. I don't shoot first and then check later if the shots were strong enough. This is what I meant when I mentioned the issue of disciplined creativity. Visual control does not rely on the camera; it lies in the person behind the camera. This is complicated. That's exactly why I said we have to be vigilant about our choice of shots. When we pick up a camera for a documentary or a narrative film, we have to ask ourselves if we are really being creative. What is it that you want to convey? Is it possible to convince an audience with your visual angles? The story alone is unlikely to move the audience. Not that many original stories are left untold.

Jia: For documentary filmmaking, I agree that the shooting distance is very important. This distance defines the relationship between the shooter and the subject. I like to stop at a certain distance and not interact or speak with them. I seldom use interview techniques to draw attention to the camera.

This is my own aesthetic style. I feel that people can express a wealth of information from mere appearances.

It's not necessary to delve too deeply into the concrete details of characters' lives. We can focus on their state of mind, their frown, their silence, the way they smoke or move. From a distance, the audience relates their own life experience to empathize with the face they see on screen. These shots are very meaningful—images can stir people's memories. If you are very passive, not ready to bring your own memories or emotions to what is on screen, you will probably find my films very boring; but if you possess what it takes and are able to use your own feelings to engage with what I want to communicate, you will find my films fulfilling.

THE POWER OF VISUAL REALIZATION

Tsai: Frankly speaking, I don't think that there is a clear line between documentary and narrative films. I don't see it. I don't think we should get stuck on those labels. Both are evolving and open to breakthroughs. Images should be challenging, and they should keep breaking new ground. They may function differently, but they should affect their audience in the same way.

Jia: I was never very conscious of form when I was shooting documentaries. They felt like narratives because I always add a lot of subjective input to the camera.

The reality captured on film is an artistic reality; it's not literal. Every shot I took served this aesthetic reality, and I took liberties. But it's interesting that I opted to make narrative films about subjects I spent a

IT'S A NARRATIVE AS WELL AS A DOCUMENTARY

long time observing. For them I wrote scripts and stories and used actors, but for subjects I was not able to live near and observe for a long time, I used the documentary form. By documenting these experiences, I access and understand different people or objects.

I consciously put *Xiao Wu*, *Platform*, and *Unknown Pleasures* together as my "Hometown Trilogy." *Xiao Wu* was made in 1997, when the economic reforms began. *Platform* takes place between 1979 and 1990 and documented all the changes China went through. *Platform* should really be the first of the three, *Xiao Wu* the second, and *Unknown Pleasures* the final chapter, since it was set in 2000. Together, they portray the daily lives of ordinary Chinese people in a small- to medium-sized city. This arrangement was inspired by a documentary mentality and was only possible through the documentary aesthetic.

Tsai: I have also been contemplating the essence of my films. Like why have I used the actor Lee Kang-sheng in all my films? My relationship with Lee is different from other lasting director-actor teams. It's not really like the relationship between Truffaut and Jean-Pierre Léaud, for example, even though some people compare us to them.

I use Lee as if to document him as a person. Every time I work on my protagonist's character, I create certain elements that resemble Lee, and then I get him to play the character. I remember the day after Wang Mo-lin[4] saw Lee's *Help Me Eros* (2007), he told Lee

---
4. Wang Mo-lin, b. 1949 in Taiwan, is a stage impresario, writer, director, and critic.

"You are such a lucky actor! I have watched you on screen since *Rebels of the Neon God* when you were in your twenties, and ever since then I've been watching you way into your thirties. You have been changing and your changes have been documented." That's the way it is. Although Lee was playing different characters, much of his real self was embedded in these characters.

Gradually, I came to realize that the differentiation between "reality" and "film" is getting blurred in my mind. In documentaries, nobody lets you shoot every moment of their life. But in a narrative film, an actor can fill in the blanks. The images are very precise; all you need to do is direct the actor's performance. What really is reality? I think it's unclear; that's why I find the medium of film so powerful.

This relates to why I tried to record the detailed changes in my actor, Lee. That for me is a quest for "reality." I couldn't really capture "real reality," but at least I could let you feel the real changes in Lee: his age, his physique, and so forth. This creative freedom is my reward in the filmmaking business. The films he's in are simultaneously documentaries and narrative films. Lee's characters and performances expose the gray area between the two forms. Strict definitions of those forms are not so important.

Jia: I genuinely admire your long collaboration with Lee. Your mutual trust gradually developed to the point that you could join forces to break taboos and really affect society. You were very bold in *The Wayward*

Cloud. Your depiction of human isolation and sexual desire, particularly in Chinese culture, was a window into the depth of the human psyche.

I said earlier that I usually don't feel much difference shooting documentaries over narratives, and that's also true the other way around. Sometimes I even want my films to be literal documents. When people watch *Xiao Wu* all these years later, they hear the sounds of 1997 China. I think this helps my films stand the test of time.

On the other hand, I often use filmmaking to observe people like a documentary—to draw out their charisma and beauty. When I filmed Wang Hongwei in *Xiao Wu*, I was basically documenting Wang's own natural movements. *Platform* and *Still Life* became portraits of Zhao Tao. She goes from her early to late twenties, maturing through her characters. She embodies a Chinese woman changing through time. This is a kind of documentary.

Tsai: To sum it up, it really doesn't matter if a work is a documentary or a narrative. People don't just focus on the images; they have way too much of an agenda on their mind. I'd rather be thinking about what images are powerful enough to affect people. The power of images is being wasted these days; that, to me, is the biggest regret.

Originally published in *Elite Reader* (*Chengpin haodu*), January, 2008.

PART III
THE DIRECTOR AS FILM AND CULTURAL CRITIC

SUMMER IN TOKYO (1999)

20

Translated by Claire Huot

On my arrival in Tokyo, Shozo Ichiyama offered to take me to Kamakura. Kamakura, a one-hour drive from Tokyo, is the site of Yasujirō Ozu's grave at the old Engaku-ji monastery. My host, Shozo, is the producer of *Flowers of Shanghai*. He heard from Hou Hsiao-hsien that I like Ozu's films, so he arranged a visit to the master's grave.

We were accompanied by Keiko Araki and Asako Fujioka. Keiko is the president of Japan's PIA Film Festival; Asako Fujioka is a member of the selection committee for the Yamagata International Documentary Film Festival. The four of us have known each other for many years. When we meet it's always abroad and always rushed. Mostly we communicate with each other via e-mail. We're never sure what the others are actually working on at any given moment.

July is Japan's rainy season, but that day the rain was light. The four of us offered fresh flowers to Ozu. We stood a while in front of his grave, which is marked only by the word "nothingness" (*mu/wu*) and then returned to Tokyo. By chance, the PIA Film Festival was opening that day, and Keiko found time to take me while we continued to talk about her work. Long ago I'd heard that the PIA Film Festival had a huge impact on Japanese cinema, and now I finally understood why.

Pia is a Japanese weekly magazine that showcases cultural events all across Japan, including films, exhibitions, concerts, dances, etc. When the magazine was founded, the editor, Hiroshi Yanai, was still a university student. At that time, Japan's economy was soaring, and Tokyo had gradually become the cultural center of the East. With the sudden abundance of cultural events, young people had no idea how to choose. Hiroshi Yanai saw an opportunity and created *Pia* magazine, which gives timely and accurate news on the arts, helping to orient young people in their consumption of culture. A mere leaflet miraculously brought him the attention of hundreds of millions of households. Still today, *Pia* is hugely influential on the art scene targeting Japanese youth. And the huge yearly revenue from the magazine's advertising has created opportunities for other activities.

Following the swift economic development of the sixties, Japan entered a period of stability in the seventies. In cinema, the Japanese New Wave had swept the country during the sixties and stimulated feverish film enthusiasm among the youth. Film quickly became an important avenue for Japan's youth to express themselves. People from diverse backgrounds and professions could grab an 8 mm camera and give expression to their inner voices. *Pia* magazine adapted to the new circumstances and, in 1977, launched the PIA Film Festival. The festival accepted films of any length and had no fixed criteria for a film's entry into the competition. It provided an opportunity for thousands of independent shorts, features, and documentaries to be screened, and it offered a venue for experimental films.

The PIA Film Festival and Japanese independent films quickly developed a fruitful relationship. Because of the existence of the PIA Film Festival, young people started to see the possibility of making their own films; even a four-minute 8 mm film might be selected by the PIA Film Festival. It might even win an award and open the world of cinema to the filmmaker. Shochiku and Toho and a great number of independent film companies saw the PIA Film Festival as a place to look for film talent. Every year a few new people are discovered by investors there, and the newcomers are given the opportunity to make a feature-length film on a larger scale. Naomi Kawase, who won the Caméra d'Or at the Cannes Film Festival for her film *Suzaku*, and Hirokazu Koreeda, who won international recognition with his films *Maborosi* and *After Life*, were among the first to be discovered and promoted by the PIA Film Festival.

Thanks to the festival, a great number of young people have taken their first steps in cinema with a minimum of financial and technical constraints. In a country with a soaring economy, an interested person can easily make a short 8 mm or digital film. What matters is that the PIA Film Festival is a channel that connects film amateurs with film companies and other film organizations. The PIA Film Festival has given countless film amateurs the confidence to take the first step, the opportunity to start small, calmly, and surefooted. Don't complain or be depressed because you can't express your talent, don't just talk like a master. Please, first shoot a short film! If you can't shoot in 35 mm, then shoot in 16 mm. If you can't shoot in 16 mm, then use 8 mm. If you can't shoot in 8 mm, then use Beta. And if you can't use Beta, then borrow a home video recorder and make a VHS. Ninety minutes is a good length but so is two minutes. Any style or format will do, the important thing is that you use your skill to show your unique view of this world.

As a result, in Japan I've rarely heard those intimidating speeches claiming that technique is everything. Nor have I witnessed people who have done well flaunting their fortunes. Filmmaking holds no mystery; it's accessible to ordinary people.

That's why Japanese cinema is in such a healthy state. It is able to face good and bad times with equanimity. Of course, the PIA Film Festival is an organization and, as such, is bound to have shortcomings. For instance, there will inevitably be partiality in the selection of films, and individual jury members' aesthetic tastes can sometimes decide a film's fate. But at least the festival provides an opportunity. An outlet, while narrow, is open to all. Last year Japan had 150 debut feature films—an incredible number.

When Keiko Araki asked me if there was a similar film festival in China, I was too humiliated to reply.

These past few years, I've witnessed too many friends having bad experiences trying to make films in China. Clutching their crumpled scripts, they come up against familiar "No Soliciting" signs. Others, not without great difficulty, manage to make the rounds between companies, but in front of all of these faces, they lose their self-respect. They become ruthless. Some of them place all their hopes in networking; they make friends with as many people as they can. They hope to find a "big brother" to come to "little brother's" assistance. But big brothers always elude them; hope is just out of reach. One day, opportunity knocks and a "boss" takes up your script only to hold on to it for a year or so. You finally discover that the "boss" is not an investor but a swindler.

Others go the route of "public relations" with foreigners. They attend a few parties in diplomatic compounds before realizing that it's not that easy to relate with foreigners. And the foreigners are no less cautious. In entertainment publications, big or small, the picture presented is always one of prosperity—you scratch my back and I'll scratch yours. But aspiring filmmakers still walk the streets of Beitaipingzhuang with heavy hearts.

SUMMER IN TOKYO (1999)

Opportunities appear numerous, but there seems to be no way to get off the ground. So they do less and less film research and they concentrate on perfecting their social skills. Sometimes they meet up with friends in the same predicament and commiserate. In the snack stall by Beihang they drink bitter alcohol and play drinking games, shouting: "We live in a world of gangsters! No one is safe from the knife! One knife, two knives…"

Indeed, cinema resembles more and more a world of gangs. Look at the way the media discuss the Fifth and the Sixth Generations—you'd think we were two opposing criminal clans. In fact, we mostly mind our own business; we don't interact. But in the world of gangs, the rumor mill keeps grinding and manufacturing resentment. Nowadays, discussions on cinema are basically gossip; people talk of film as if it were money. This has been going on for years; isn't it time to move on?

I think the experience of the PIA Film Festival is paramount for Chinese cinema to replicate. Professional fields must always find means to recruit new talent, just as a population needs a constant supply of water. New life must spring up from below— made of the experiences and hopes of the lower strata, smelling of mud. There must be a steady stream of vitality. Improving the professional part of the field is a problem of another order. You need regulations and shared standards. When those are in place, then things can be simple; then there's less need to appeal to personal feelings and interpersonal relations. Whether it's art or business, professional standards will always be more reliable than relations between people with mercurial moods or networks full of resentment.

Before leaving Tokyo, I attended a cocktail party in the Imperial Hotel to mark the closing of the PIA Film Festival. At first, the hall was filled with middle-aged people in suits and ties, which made me feel uneasy. I spotted Naomi Kawase's husband, Takenori Sento, a producer who enjoyed success early in his career. He was also wearing a Western-style suit. I complained to him that the cocktail party was stuffy. He shrugged his head helplessly. He was a jury member, and that's just the way things were. Yōichi Sai approached and invited us to attend the screening of his new film. Old Sai had shaken the Japanese film scene with his *All under the Moon*, but he looked as nervous as anyone prior to the premiere of his new film. While he was talking, a crowd of young people with dyed-blond hair, wearing sneakers and clothing draped in metal chains, burst in. The cold formality of the conference hall instantly vanished with the arrival of these young people who had just won prizes at the PIA Film Festival. That's the force of youth.

The people there were all in high spirits; it was their event. It had nothing to do with me, a Chinese guest. Just as gloomy thoughts threatened to fill my head, Asako Fujioka told me that this year's Yamagata International Documentary Film Festival had invited Lin Xudong to be a jury member. This definitely brought my spirits up. The international scene needs Chinese participation; Professor Lin has been promoting Chinese documentary films for many years, and he can offer a fresh perspective. Outside, it began to rain again in Tokyo. I suddenly wanted desperately to get back to Beijing to my own work.

Originally published in *Southern Weekly* (*Nanfang zhoumo*), August 20, 1999.

21. WHO IS USHERING CHINESE-LANGUAGE CINEMA INTO THE NEW CENTURY?

Translated by Claire Huot

Last year, after the Venice International Film Festival, Zhao Tao, the leading actress in *Platform*, and I traveled around France. We were waiting to go to the Toronto International Film Festival to promote the film. During our stay in Paris, I saw an advertisement in *Libération* for Edward Yang's (Yang Dechang) new film *Yi Yi* (*A One and a Two*). The picture showed a little boy seen from the back who is about to climb up a very tall red staircase.

Just by looking at the ad, I felt that Yang was repeating his previous films' mistakes. I can't say much for his earlier films. Even his best feature, *A Brighter Summer Day* (*Gulingjie shaonian sharen shijian*), is saturated with more or less uncontrolled feelings and ideas. Director Yang is very conceptual and I don't care for that kind of cinema.

When Zhao Tao heard that it was a Chinese-language film, she wanted to see it. I accompanied her on the crowded subway all the way to a theater close to Le Centre Pompidou, where we elbowed our way through the crowd to buy our tickets. To our surprise, there was a long queue outside the cinema; it was drizzling, yet the audience waited patiently. I was moved by the moviegoers' reverence. It suddenly occurred to me that cinema was a sacred thing. I was reminded of O. Henry's story of a vagabond who walks by a church, hears the organ playing, and is transformed.

Nevertheless I was laughing up my sleeve. Zhao Tao once told me that her favorite film is *The Lion King*, so I was sure she would not be able to withstand Yang's "philosophical film," running two hours and forty minutes. Since I was not a fan of Yang, I was prepared to make an escape in the middle of the screening.

But once the film started, I was drawn into Edward Yang's meticulously constructed mundane world. This is a film about family, midlife, and human predicaments. The story develops around Wu Nien-jen,[1] who represents the middle class, and reveals the truth beneath the surface "happiness" of a normal Chinese family. I can't recount the film's story detail by detail because the real portraits of "happiness" that pervade the film made me feel tense and heartbroken. At the end, when the small child says, "I'm only seven, but I feel old," I felt completely dejected. In this masterpiece, Edward Yang has soberly described the pressures of life. And it left me exhausted and gasping for breath. I can't connect *Yi Yi* with his other films because this time Edward Yang really surpassed himself. His life's valuable experiences were finally presented without being overshadowed by unnecessary concepts. In a slow and painful shedding process, he exposed the true face of his fifty years of life. As for me, on a rainy afternoon in Paris, I saw the most brilliant film of the year 2000.

When the theater lights came back on, I noticed that Zhao Tao's eyes were red. I would never have imagined that this young woman who loves cartoons so much would be able to watch such a long film, or that almost nobody in that packed French audience left the theater early. Everyone applauded. Facing the silver screen, with the images that had just disappeared but were

---

1. Wu Nien-jen (b. 1952) is a leading figure in Taiwanese cinema. He is an actor, screenwriter, director, and producer of numerous New Wave films.

WHO IS USHERING CHINESE-LANGUAGE CINEMA INTO THE NEW CENTURY?

still vivid, we all saw ourselves. Zhao Tao, who is a dance teacher, asked me why we didn't see such films on the Mainland: I didn't know the answer. Our cinema doesn't seek truth. Happiness is fine as a topic, because that happiness is not real.

Soon thereafter it was September and *Platform* was having its Asian premiere at the Busan International Film Festival. I went with my cinematographer, Yu Lik-wai. *In the Mood for Love* (*Fa yeung nin wa / Huayang nianhua*) was the closing film of the festival. Yu Lik-wai had also been the second unit cinematographer for *In the Mood for Love*, yet he had not seen the finished version and couldn't wait for the closing ceremony.

In the bar we happened to meet Wong Kar-wai. Behind his dark glasses he wore a wicked grin and said that he would be going to Beijing to drink with us there. As we talked some more, we learned that he was extremely proud because *In the Mood for Love* had been approved for viewing in the Mainland. I knew that he was sincerely happy: when you know your film won't be screened in the foreseeable future, you feel more than a bit disappointed. A fever for *In the Mood for Love* had taken hold of all Busan. When we came across young people carrying film posters, in eight or nine out of ten cases, the poster was for *In the Mood for Love*. I returned home before the closing ceremony. I heard that on the closing day, the weather dropped suddenly. *In the Mood for Love* was screened outside, and thousands of spectators sat in a cold wind reveling in the craze.

The force of a popular trend is tremendous. I arrived in Beijing at noon, and by the afternoon I had bought a VCD of *In the Mood for Love*. Where Wong Kar-wai's narration breaks off and actors Maggie Cheung and Tony Leung Chiu-wai walk as though dancing under a high-speed camera and frenetic music, I suddenly connected this technique to the poems that serve as transitions between chapters in traditional Chinese novels. It turns out that director Wong knows China's ancient popular culture well; his alternation between tense and relaxed scenes stems from the foundations of Chinese traditional culture. One can't say it's the cheongsam dresses or illicit love affairs that attracted middle-aged viewers; rather, it's the atmosphere that Wong Kar-wai created. That atmosphere is why middle-aged viewers were also swept up in the craze.

By the time I went back to Paris, it was the end of October, and the poster for *Crouching Tiger, Hidden Dragon* (*Wohu canglong*) was plastered all over Parisian metro stations. In the public square of the Hôtel de Ville stood a wall-sized screen, on which the film's actors Chow Yun-fat and Zhang Ziyi were flying to-and-fro in a bamboo grove, mesmerizing the passersby. I guess they were considering their knowledge of mechanics and pondering how Chinese people could overcome gravity. This was the publicity for *Crouching Tiger, Hidden Dragon*, astutely abbreviated to *Dragon and Tiger* by the French film distributor. I had started watching martial arts films from Taiwan and Hong Kong as of my second year of junior high. I knew this fantasy world from watching King Hu's 1960s films such as *Raining in the Mountain* and *A Touch of Zen*. But the mysterious Orient continues to fascinate the public. Before the United States had released the film, friends from New York were phoning and asking me to send them a pirated copy of *Dragon and Tiger*.

A few days later, I saw Ang Lee in London, exhausted from the worldwide success of the film. We chatted in a bar surrounded by Warhol works hanging on the walls. I worried that the air conditioning would blow them down. As we discussed *Crouching Tiger, Hidden Dragon*, Ang Lee said this: "Don't think about what the audience wants to see; you should think about what they haven't seen." I took this statement as a sign of Ang Lee's business flair; I committed it to memory.

The films by Edward Yang, Wong Kar-wai, and Ang Lee respectively represent three directions in creation: Edward Yang depicts life's trials and tribulations, Wong Kar-wai launches fashionable trends, and Ang Lee produces blockbusters. These three creative directions highlight the fact that Chinese-language films made with different production models hold immeasurable creative energy, and they demonstrate the healthy state of Chinese cinema. Today there is no need to reiterate the successes of these three films. In France, *Yi Yi*'s audiences exceeded 300,000 at the box office; *In the Mood for Love* attracted 600,000, and *Crouching Tiger, Hidden Dragon* achieved the highest number, with 1.8 million tickets sold. Those who understand how cinema works must realize that this is practically a miracle. This miracle has put Chinese-language cinema back on the map. These films have resuscitated Chinese cinema, which was slowly declining into a dismal state. I myself am benefiting from the favorable context these three filmmakers have provided: *Platform* sold relatively well, which means that it will reach a broad audience.

Having said this, it is hard not to notice that of the three directors, two are from Taiwan, and one is from Hong Kong. If cinema is an embodiment of culture, it would appear that the vast Mainland has already sunk and that the heroes who have rescued Chinese-language cinema come from humid little outer islands. From the mid-1990s onward, our domestic films have lost creativity and credibility in the international market. In fact, already a few years before that, our big-time international film directors had almost no international distribution; they relied on media hype to boost appearances. Those directors who walk around believing they have an audience put on a Beijing accent and follow in the steps of the delusional Ah Q.[2] A cinematic desert lies in front of us. My view on this? There's nothing to view.[3]

We stood by as the three filmmakers—Edward Yang, Wong Kar-wai, and Ang Lee—ushered Chinese-language cinema into the new century. The absence of Chinese Mainland filmmakers in their midst has apparently not provoked restlessness among Mainland film-workers. Their complacency has convinced me that a new era must begin now.

Originally published in *Southern Weekly* (*Nanfang zhoumo*), February 16, 2001.

---

2. Ah Q, the character created in the 1920s by China's most prominent twentieth-century writer, Lu Xun, has come to represent for the Chinese the flaws in the Chinese national character, such as petty subjectivity and stubborn ignorance.

3. Jia is playing with the word *kan*, which can mean, when grouped with *wo* in an interrogation (*Wo kan*?) two things: "My opinion on this?" or "Do I watch [their films]?"

THE UNSTOPPABLE MOVEMENT OF IMAGES: NEW FILMS IN CHINA SINCE 1995

22

Translated by Sebastian Veg and Claire Huot

In 2001 the *Beijing Evening News* reported that a vendor of pirated DVDs was arrested in the office of the literature department of the Beijing Film Academy. This person came to the academy regularly to sell DVDs to teachers and students and was denounced by some students with strong feelings about intellectual property rights. The literature department was also drawn into the case, since it had provided him a place to sell his DVDs. The unofficial version of the story was even more colorful: the vendor had been denounced not by a student but by a colleague engaged in the same pirate trade. The whole thing sounded like the plot of a gangster film, and what made it even more absurd was that it had happened at the famous Beijing Film Academy.

At most, this trader in pirated films was a small-time retailer; but the story of his fortune, as spread by the students of the Film Academy, sounded like the plot of an inspirational movie.[1] In 1999, when DVDs were just becoming common in China, this man, who had come to Beijing from the provinces, would trek to the Film Academy every day with a satchel of DVDs to ply his trade. After half a year, he bought a motorcycle, which allowed him to shuttle between Beijing's numerous colleges and universities. In 2000 he exchanged it for a second-hand jeep, and at the

---

1. Inspirational or inspiring movies (*lizhipian*) refer, in China, to feel-good movies such as *Forrest Gump* and *The Shawshank Redemption*.

time of his arrest, people noticed that he had just bought a brand new Volkswagen Santana sedan.

Of course, from an economic and legal standpoint, this story demonstrates that China still has a serious problem with intellectual property rights, but in terms of culture it shows the huge demand for films in China. Although a yearly average of 250 official films were released during the peak of film production in China, international films, including many influential classics, were not available. Before pirated DVDs began circulating, it was unimaginable that an ordinary citizen could see films like Godard's *Breathless* or Tarkovsky's *The Mirror*. Even popular American films like *The Godfather* or *Taxi Driver* were hard to find. People had little knowledge of movies and no way of sharing the cultural experience of cinema that had developed over a century.

It's not difficult to understand the lack of moving images in China before Deng Xiaoping's "Reform and Opening Up" policy of 1979. In the preceding era, China was in a cultural straightjacket. Apart from so-called revolutionary art, there was not a new opening for any other form of culture. But to understand what happened after the end of the Cultural Revolution, especially the limitations imposed on film in China from the 1980s onward, it is necessary to take a slight detour.

Following the thaw in national politics after the Cultural Revolution, ideas began to disseminate more freely. Great quantities of modern Western literature, music, and art began to be translated and introduced to the Chinese people. By 1989, there had even developed a nationwide enthusiasm for philosophy: works by Nietzsche, Sartre, Freud, and others were printed and reprinted, and the Ministry of Culture began to commission the translation of works by all the Nobel Prize laureates for literature. After Deng Xiaoping's visit to the United States, American country music began to be heard on the radio, and many people, wearing their first pair of jeans, walked around humming "Leaving on a Jet Plane."[2]

The film world, too, was crying out for reform. In 1979, Zhang Nuanxin, director of *The Seagull* (*Sha ou*), and her husband, the writer Li Tuo, together published the article "On the Modernization of Film Language." They introduced André Bazin's aesthetic theories and proposed the study of Western cinema as a means of changing Chinese cinema's outdated formal language. The article provoked a theoretical renaissance, and film criticism became an important instrument in the intellectual and cultural modernization that occurred through the 1980s. Literary and philosophical concepts, such as "stream of consciousness" and "defamiliarization," were disseminated thanks largely to debates on Western cinema. Chinese intellectuals were inspired by such films as Alain Resnais's *Last Year in Marienbad* and Michelangelo Antonioni's *Red Desert*, and terms like "Godard," "jump cut," "Left Bank," and "Fassbinder" appeared more and more often in articles.

Free access to films like *The 400 Blows* and *Nights of Cabiria*, and to novels such as *One Hundred Years of Solitude* and *The Metamorphosis*, was a first step toward the enrichment of Chinese intellectual life. However, contrary to expectations, the cultural authorities, faced with this overpowering tide of ideological liberalization, continued to strictly control films coming in from abroad. Only a tiny number of professional film workers and elite intellectuals were allowed to see these films, and only at restricted screenings closed to the general public. These viewings were called "internal screenings," and the films were labeled

---

2. In 1992, John Denver toured China.

"internal reference films." During the long period of the Cultural Revolution, the only films authorized for screening were Beijing Opera movies: the eight "model operas" presenting heroic stories of the Party and the Army intended to educate the general population. The name "internal reference film" gave the restricted screenings a strong academic or professional cachet, making it clear to the public that these films could only be viewed "internally" with exclusive permission. In the early 1980s, in response to the public's, and especially intellectuals', demand for international films, cultural authorities tried to extend the screenings of internal reference films. Their approach was twofold: they enlarged the circle of those with privileged access while simultaneously reinforcing the ban on public access.

In this way, the privilege of viewing internal reference films was extended from fewer than one thousand officials and political cadres to a wider stratum of elite intellectuals. Cultural work units in Beijing and Shanghai received permission to organize occasional internal screenings with the aim of satisfying the academic needs of a small number of professionals. Free tickets were distributed. There was only one problem: although the group of privileged people had been extended, viewing films remained a privilege. For example, cities outside Beijing and Shanghai were mostly unable to organize similar screenings for intellectuals, and, apart from a small number of so-called professionals, ordinary citizens were turned away at the theater doors. Nevertheless, the exciting atmosphere of the screenings carried over into the media: intellectuals delighted in discussing *The Deer Hunter* or *Kramer vs. Kramer*, and the names of Meryl Streep and Dustin Hoffman began to reach Chinese ears. Thus, an unsatisfactory situation was covered with a thin veneer of culture.

THE UNSTOPPABLE MOVEMENT OF IMAGES: NEW FILMS IN CHINA SINCE 1995

This brings up the question: Why was a relatively liberal policy adopted toward literature and music while film was singled out for tight control? Lenin is supposed to have said, "Of all the arts, for us, cinema is the most important." Film was seen as endowed with a greater capacity to influence because of its ability to overcome the obstacle of language. Even the illiterate can understand new ideas through images. That's why, beginning in the 1980s, people were free to buy Proust's *Remembrance of Things Past* in a bookstore, but there were no public screenings of any of Godard's films. Steadfastly following Lenin's teachings, cultural officials censored access to moving images. Even in today's China, cultural policy is by no means as free as economic policy, and film has the misfortune of being the most conservative link in the cultural chain.

The situation only began to slowly change after 1995, but no one imagined that what would loosen the controls over film would be pirated VCDs from the coastal areas of Guangdong and Fujian. The fishermen journeying along the coasts in their wooden boats had early on supplied their compatriots with Sony tape recorders and pop music from Hong Kong and Taiwan. Now, fifteen years later, they used the same method to smuggle films from Hong Kong and Taiwan. Before VCDs became widespread, the expense of video recorders and VHS tapes made home viewing of films a luxury. But in 1995, more than a dozen factories producing VCD players sprang up in southern cities, and fierce competition rapidly brought the price of VCD players down from over RMB ¥3,000 to around RMB ¥800 [from USD $480 to around $128]. Most urban residents were now able to afford electronics, and installing a "home cinema" soon became the fashion. A simple system composed of a television set, a VCD player, an amplifier, and two speakers was all a middle-school teacher or taxi driver needed to hold private screenings at home. This was also the time when computers were entering people's lives, and installing a CD drive on one's PC became extremely popular.

At first, the films smuggled back on fishing boats from Taiwan and Hong Kong were a startlingly diverse but haphazard mix: a stack of Hong Kong martial arts films from the 1970s might also contain a copy of *Citizen Kane* or *Battleship Potemkin*. But they were still mostly commercial Hong Kong films: John Woo's masterpieces, *A Better Tomorrow* (*Yingxiong bense*) and *The Killer*, Tsui Hark's *Once Upon a Time in China* (*Huang Feihong*), and Jackie Chan's kung-fu comedies were all received with enthusiasm. A wave of American films arrived soon afterward and smugglers kept up with the newest trends in Hollywood while also providing their customers with retrospectives of American cinema. For example, when *True Lies* was released in the United States, a pirated version was available in Beijing one week later. These VCDs distributed almost simultaneously with Hollywood were invariably "gunpoint versions," shot in theaters with a handheld camera aimed at the screen and immediately copied. In these versions, you could often hear the audience laughing or coughing and sometimes an audience member leaving in the middle of the film would even float across the picture. Films such as *The Godfather*, *Taxi Driver*, and *The Graduate* were pirated one after the other, and people began to pay attention to the tastes of intellectuals. This is when small shops specializing in European films began to appear on the streets. To the great displeasure of the French broadcasting group MK2, one of the first of these pirated films was one of Kieslowski's *Three Colors, Blue*. Juliette Binoche gained a million fans.

To the vast majority of Chinese audiences, even a film shot in 1910 was new. Famous films, directors known only by name, and actors seen only in film magazines were suddenly available to ordinary people, engendering an irrepressible enthusiasm. Official control over films was effectively reduced to control over the legal sale of films. The Film Bureau resorted to diplomatic channels to express its deep discontent over *Red Corner* starring Richard Gere, but the pirated VCD of the film could be viewed throughout Beijing. Sellers of pirated copies established a network covering all of China's urban and rural areas, and it is probably no exaggeration to say that wherever there was a post office and a gas station, there would be a local retailer of pirated VCDs. Chinese people, who had been deprived of almost all images, gained the right to watch movies freely thanks to pirated discs. This all adds up to an embarrassing predicament in which we had to overcome an irrational policy of film isolationism by engaging in illegal pirating.

Most consumers of pirated films believe that the act of pirating is wrong, but nobody feels guilty buying pirated discs. People were caught up in the double euphoria brought on by the restoration of rights and the actual viewing of films. In 1999, similarly inexpensive domestic-made DVD players flooded the market, and people began updating their equipment to watch DVDs with a higher audiovisual quality. As if to match this technological progress, sellers of pirated discs also began to offer better films than in the VCD era. Fellini, Antonioni, Tarkovsky, Godard, Rohmer, Kurosawa, Hou Hsiao-hsien: copies of all the most important films in the history of cinema seemed to be available to all.

In 1997,[3] the first unofficial film club, Office 101 (101 Bangongshi), was established in Shanghai. It demonstrated that

---

3. An online book review of Jia's original book gives the date of the founding of Office 101 as October 1, 1996. "Must-Reads: The Ten Best Chinese-Language Books of 2009 on Cinema," December 24, 2009, http://group.mtime.com/12119/discussion/780125/5/.

young people were no longer satisfied with private screenings and that they yearned to establish a platform for free expression and exchange. Office 101 had about two hundred regular members, including teachers, workers, office employees, and students. The initiator, Xu Yuan, was a former employee at the Shanghai Customs office who resigned on founding the club. The organizers chose the Hongkou Cultural Palace as their venue for regular screenings. The screenings were followed by open discussions, the content of which was reported in their self-published unofficial journal. After 1999, similar film clubs appeared in great numbers. In Guangzhou, the Southern Film Forum (Nanfang dianying luntan) was organized around a group of journalists and people working in audiovisual media: Rear Window Film Viewings (Houchuang kan dianying) was created in Nanjing; Wuhan Film Watchers (Wuhan guan ying) in Wuhan; People's Cinema (Pingming dianying) in Jinan; Hands-On Club (Shijian she) and Fanhall Films (Xianxiang gongzuoshi) in Beijing; Free Film (Ziyou dianying) in Shenyang; and Changchun Film Study Group (Changchun dianying xuexi xiaozu) in Changchun. Leafing through an atlas of China, you would find cities with film clubs all across the Mainland.

The more surprising thing was that not long after having been established, most of these clubs extended their activities to a broader public. In Beijing, the Hands-On Club used the Box Bar (Hezi) near Tsinghua University as a screening venue and regularly organized specialized screenings, such as retrospectives of Fellini or Luis Buñuel. At the outset, only members were internally informed of the planned activities, but very soon information about the screenings appeared in the pages of *Beijing Youth Weekly* (*Beijing qingnian zhoukan*), the *Guide to Quality Shopping* (*Jingpin gouwu zhinan*), and other popular

THE UNSTOPPABLE MOVEMENT OF IMAGES: NEW FILMS IN CHINA SINCE 1995

publications. Rapidly, almost all the groups set up their own BBS on the Internet, seeking to open up a platform for discussion. The most famous film discussion forum was the Nanjing BBS run by Rear Window Film Viewings, while the Hands-On Club's Yellow Pavilion Images Online (Huang tingzi yingxiang) also boasted a large number of hits.

Through www.xici.net, it was easy to find the homepages of all these clubs. On their BBSs, discussion moved from the initial critiques of Western films to reflections on practical questions confronting Chinese cinema. Independent Chinese films, which had been shot since the 1990s but had never been publicly released, began to draw attention. Zhang Yuan's *Sons* (*Erzi*), Wang Xiaoshuai's *The Days*, He Jianjun's *Postman* (*Youchai*), Lou Ye's *Weekend Lover* (*Zhoumo qingren*)—all these films that had lain covered in dust began to appear on the program lists of the film clubs. Once these previously banned films had been projected in the noisy environment of a bar and onto people's retinas, it became clear that they were very much alive and could no longer be suppressed.

Zhang Yuan, Wang Xiaoshuai, and the first cohort of independent Chinese directors began working as rebels against the system. In 1989, they had just graduated from the Beijing Film Academy. After the great social unrest,[4] a number of intellectuals began to adopt a way of life that set them apart from the system. These feelings are manifested in the anger of *Beijing Bastards* and the loneliness of *The Days*. Movies no longer followed the production techniques associated with the studio system. The feelings depicted were at odds with the established film system, and inevitably these films were banned by the authorities. However, the independent film movement that began under

---

4. Jia is referring to the Tiananmen Incident (June 4, 1989).

these circumstances remained largely misunderstood by the public for a long time. It was only when the film club organizers borrowed VHS tapes directly from the directors and showed their films in bars that the notion of independent film began to be understood by more and more young people.

Today, more young people from outside the film system are preparing themselves to shoot independent films. Just as reading can arouse the desire to write, the uncensored viewing that began in 1995 has instilled an interest in shooting films in a growing number of young people. Some of them are students, while others are workers, office employees, writers, or poets. The technical limitations they face because of their lack of expertise are greatly reduced in this vibrant era of digital video.

The change brought about by DV in China is no less than a change in the cultural habits of an entire nation. Before DV, the Chinese people lacked a tradition of capturing images to express themselves, to the degree that few people even took photographs. Reading and writing literature were the mode of expression we excelled in, while visual experience was sorely lacking. After 1949, the government decided that only official film studios would have the right to shoot films, and cinema effectively became an artistic monopoly. We were isolated from film for so long that we forgot that expressing ourselves through film was always our right. When the Chinese people began trying to look at the world through a viewfinder, DV provided not only a new mode of expression but also a restoration of their rights. After 1995, the majority of experimental film directors chose to abandon the path of making films within the censorship system. Working in an unofficial manner and taking an independent stance, Chinese people began creating a brand new film world outside the system and progressively organizing their own visual culture.

THE UNSTOPPABLE MOVEMENT OF IMAGES: NEW FILMS IN CHINA SINCE 1995

Many directors using handheld DV cameras chose documentary film as a starting point for their creative work. This stimulated further enthusiasm, because in the century-long history of Chinese film, two traditions have been absent: documentary film and experimental film. The trend of making independent, unofficial films made possible by DV compensated for this lack in Chinese cinema. These new directors were even more widely geographically distributed than before. In the past, the majority of directors lived in the cultural capitals of Beijing and Shanghai, but now even places as far away as Sichuan or Guizhou have filmmaking fever. The best known among these directors are Yang Tianyi, Du Haibin, and Wang Bing in Beijing, as well as Ying Weiwei in Shenyang, Hu Shu in Guizhou, and Wang Fen in Jiangxi.

In 1996, Yang Tianyi—at the time still employed as an actress in the Drama Troupe of the People's Liberation Army's General Political Department—began to shoot her first documentary, *Old Men* (*Lao tou*). Her film came about quite fortuitously: Yang Tianyi would frequently walk by a row of old men basking in the sun in her neighborhood. She was moved by this scene and began shooting their lives using her handheld DV camera. In the following two years, during which she continuously filmed them in the slow passage of time, she captured their attachment to and reverence for life, documenting the end of their lives and their farewell from existence. Du Haibin, a graduate of the Beijing Film Academy's photography department, having returned to his hometown of Baoji in Shaanxi Province for the Spring Festival of 2000, met a group of vagabond youth along the railway tracks. He shot *Along the Railway* (*Tielu yanxian*) about them. Drifting along with them, his camera cut through the surface of their miserable lives and entered their youthful dreams.

Wang Bing's *West of the Tracks* (*Tiexi qu*) is a kind of super-production in which the author edited three hundred hours of material he had shot over two years to craft a trilogy of independent but spatially related films. Tiexi is an industrial district in the northeastern city of Shenyang, where state-owned factories are currently facing bankruptcy, leaving the whole area in hopeless desolation. The real situation of workers appears even grimmer in the icy weather of the Northeast. Areas like Yanfen Street are slated for destruction; old men in the workers' sanatorium fish for tiny shrimp by scooping them up with plastic bags; families stare blankly at their television sets, while the old train keeps rumbling through Tiexi District. All these scenes compose a panorama of the failure of the planned economy.

Director Ying Weiwei lives in Shenyang, but I don't know whether she is familiar with Tiexi District. Her background is in Chinese language and literature, and her film *The Box* (*Hezi*) documents the life of a lesbian couple living together in an apartment unit. During their uninterrupted heart-to-heart talks, we watch them by turns hurt and care for each other. The documentary form thus penetrated a hitherto sealed world to open up an unfamiliar private space. Another female director, Wang Fen, turned her camera on in her hometown of Jinxian, Jiangxi Province, in the year 2000. She filmed her parents, revealing family secrets, endlessly repeated recriminations, and inescapable family obligations. This film, entitled *More than One Is Unhappy* (*Bu kuaile buzhi yige*),[5] is about the collapse of family ties. Both these works have the feel of Japanese "I-novels" (*watakushi shosetsu*),[6] demon-

---

5. The film won the New Wave Prize at the Yamagata International Documentary Film Festival in 2001.

6. The I-novels constitute a genre in modern Japanese literature. Often narrated in the first person perspective, events in the novel refer to the author's life. These works typically depict the darker, seedier, and scandalous sides of life.

strating the continual dissolution of ideology in film.

Hu Shu, from Guizhou, works for an official television channel but made *Leave Me Alone* (*Wo bu yao ni guan*) independently. In Guiyang, he followed the sentimental lives of several "three-services escort girls" (*sanpei xiaojie*),[7] watching them fall in love and confront betrayal. Their thirst for love makes their youth in a quiet border town seem painfully melancholic. One cannot help but be reminded of Shen Congwen's light and informal style steeped in the pathos of life or the Chinese film classic *Spring in a Small Town*, made in 1947.

Another director of note, Zhong Hua, two years after leaving the army, devised a way to go back to the garrison where he had been stationed. He independently shot a documentary about the lives of active servicemen, entitled *This Winter* (*Jin nian dong tian*).[8] Directors with strong links to the "system" but who willingly distance themselves from mainstream cultural practice have become a quite common phenomenon in recent years. Documentary film has enriched film aesthetics throughout the country and has turned the hands of the cinema's clock to the present. They look at real-life conditions to tell the story of the Chinese people's contemporary experience. Documentary is thus breaking with the theater-based tradition that has always dominated Chinese cinema.

Among the first to make experimental films were Guangzhou Academy of Fine Arts graduate Cao Fei and Shanghai artist Yang Fudong. Cao Fei's earliest work, *Chain* (*Lian*), is strongly reminiscent of Buñuel's *Un chien andalou*: among the fragments

---

7. A common euphemism for prostitution: in theory the three services are chatting, dancing, and drinking with the client.

8. The film was awarded the Prix Premier in the International Documentary Film Festival of Marseille in 2002, the year in which Jia Zhangke's *In Public* received the top prize in the International Competition.

of cut-up film clips you occasionally glimpse a simulated surgical operation. Machines, body parts, and blood from an unseen source, along with bridal veils and plastic flowers, seem to express a feminine anxiety. Yang Fudong's work *Hey, Sun Is Rising! (Hei, tian liang le)* reminds the viewer of the uncanny atmosphere in Philip Kuhn's book *Soulstealers: The Chinese Sorcery Scare of 1768*.[9] Fleeting spirits wander along broad southern streets grasping swords in a kind of Chinese-style restlessness. This year, Yang Fudong presented his new ninety-minute black-and-white film *An Estranged Paradise (Mosheng tiantang)* at the Kassel biennale. In the film, a poet suddenly thinks he has fallen ill, but after all sorts of tests, his spirits rise again. This happens during the Plum Ripening rainy season in the southern city of Hangzhou, which is sometimes called "Paradise." A number of other film works, perhaps more correctly classified as "video art," have been exhibited as installations in art museums.

Of course, there are far more people shooting feature films than documentaries or experimental films. Nanjing director Zhu Wen was one of the first to shoot feature films using DV. Before that, his novels had already attained a tremendous following among young people. He helped Zhang Yuan complete the script for *Seventeen Years (Guo nian huijia)* and wrote the script for *Rainclouds over Wushan (Wushan yunyu)* for Zhang Ming. In the winter of 2001, he shot the film *Seafood (Haixian)* in a little over ten days in the northern seaside resort of Beidaihe. It tells the story of a "three-services escort girl" who wants to commit suicide and a policeman who will not let her die. The policeman uses his power to obtain sex, but the young woman, after

---

9. Philip Kuhn, *Soulstealers: The Chinese Sorcery Scare of 1768* (Cambridge, MA: Harvard University Press, 1990), and published in Chinese translation by Sanlian in 1999.

killing the policeman with a gun, decides to continue living. The word *"haixian"* (seafood) connotes rawness in contemporary Mandarin. This film is a crude depiction of a tough reality. In Shanghai too, people are creating independent productions: the young director Chen Yusu collaborated with best-selling author Mian Mian to make *We're Scared (Women Haipa)*, describing the lives of young people in this city that have reinstated the spirit of capitalism.

Last summer, a young man from that city actually showed up at my office in Beijing. He was a student from Shanghai University who was planning to hold a series of screenings on campus. They wanted to show *Xiao Wu* and asked me for a high-quality video recording. I took a disc out of the drawer and gave it to him. In exchange, he took out a contract and asked me if I would agree to sign it. It was a letter of authorization, stating that the holder of the copyright for *Xiao Wu* in China, Jia Zhangke, agreed to have this film screened within the stated event. This was the first time I had seen anyone in China who understood how to deal with copyright. I happily signed my name, although it said at the top that the fee they would pay was equal to zero yuan [RMB].

China is developing at full speed; everything is moving very fast. For us, the most important thing is to hold tightly to our cameras and just as tightly to our rights.

Originally published in *Cahiers du cinéma*, n.d.

I HEARD THE SPRING OF CINEMA IS UPON US   23

Translated by Claire Huot

In 2002, director Tian Zhuangzhuang did a remake of Fei Mu's film *Spring in a Small Town* (*Xiao cheng zhi chun*) (1947). This rekindled people's enthusiasm for domestic films. I had bought my ticket, entered the Youth Palace Cinema, and sat down before I realized that I hadn't seen a Chinese film in over a year.

The newly produced *Spring in a Small Town* hardly resembles most remakes, which recklessly amend the original work in order to show the director's new understanding of the subject. Surprisingly, Tian Zhuangzhuang's film is thoroughly permeated with genuine respect for the original work, which is gratifying. Looking back at China's film history, it's clear that our film tradition has undergone too many ruptures. Now there was someone delving into the recesses of history and paying tribute to an oft-misunderstood tradition. It requires a tremendous sense of cultural duty to undertake such a task. If we add to this Tian's support of younger directors, there can be no question that he has taken on a huge responsibility. Above, he faces tradition, while below, he guides and supports newcomers. This is like a family's hardest worker.

Both Chen Kaige and Zhang Yimou also produced new work this year. Both their films, *Together* (*He ni zai yi qi*) and *Hero* (*Yingxiong*), aim to revitalize China's mainstream film industry, an industry that is seriously worrying us. *Together* applies a one-time vigorous CPR to the industry; *Hero* resembles more an emergency defibrillator. As first aid responders, Chen Kaige and Zhang Yimou seem competent enough, but what's surprising is that, taking on their new role, the two directors have begun to renege on the values they used to represent. Chen Kaige confesses that his films about art are purely selfish, while Zhang Yimou enthusiastically identifies with the values of Hollywood. I'm quite disappointed to see that *Yellow Earth* (*Huang tudi*) (1984) and *The Story of Qiu Ju* (*Qiu Ju da guansi*) (1992) have been consigned to the dustbin so quickly.

Among the new filmmakers, there's Lu Chuan, a young director whose graduate thesis was entitled "A Filmmaker inside the System."[1] Lu Chuan is interested in making genre films, and his 2002 film *The Missing Gun* (*Xun qiang*), fulfills that goal. *The Missing Gun* is also infused with the director's sense of social responsibility, which is laudable.

This year, Wang Bing completed his nine-hour documentary *Tie Xi Qu West of the Tracks* (*Tiexi qu*). This independent film of epic proportions presents the wounded face of a city district (Shenyang) after the collapse of the planned economy. The documentary marks an important milestone for recent Chinese cinema, something only an independent production could have accomplished.

From the northern city of Harbin, Korean minority director Jin Shangzhe gave us *It's Real Cold Here* (*Zan zher zhen leng*). From Guangzhou in the south, we got Gan Xiao'er's *The Only Sons* (*Shan qing shui xiu*). Andrew Cheng Yusu shot *Welcome to Destination Shanghai* (*Mudidi Shanghai*) in Shanghai's red light district. Initially, independent filmmakers were mainly

---

[1]. Lu Chuan's thesis was on Francis Ford Coppola. Its subtitle was "A Study of Coppola under New Hollywood." Lu admires independent-minded filmmakers who work from within the system, such as Coppola and Joel Coen, who visited him in Beijing in 2012.

concentrated in Beijing, but gradually that situation has changed because, from north to south, the idea of making independent films has become increasingly popular. Perhaps we have really reached a golden age of cinema.

What's different from the past is that, today, even people outside the film business have begun to learn about censorship issues. Newspapers discuss film-rating systems and why it will be so difficult to reform a system that everyone agrees is illogical and has impeded the industry's development, as well as our culture.[2] Now, at last, the issue is out in the open. Perhaps that's progress. A few days ago, a friend phoned me to tell me the ban on Huang Jianzhong's film *Rice* (*Da hong mi dian*) might be lifted. He spoke for a long time and then abruptly fell silent. After a pause, he said, "I heard the spring of cinema is upon us."

I'm not entirely sure I should trust news about the coming of spring heard over the phone.

This text first appeared as a preface to the 2003 special issue of the Third Edition of the Chinese Film Media Awards (*Huayu dianying chuanmei dajiang*).

---

2. China does not yet have a film-rating system, but a project is in the works to develop it. Presently, a board of censors decides which films are acceptable.

THE WORLD SITS ON A TATAMI 24

Translated by Claire Huot

Yasujirō Ozu was born and died on the twelfth day of the twelfth month. Birth and death fell on the same day, as though the heavens wanted to paint his destiny in Asian colors, in accordance with Buddhism's wheel of life and the fated lives in Ozu's own films. In Asian terms, a full cycle is sixty years. Ozu died on the day of his sixtieth birthday, exactly one *jiazi*, a complete life cycle.[1]

Life and death are so arcane, yet Ozu's films are ensconced in the most mundane existence. This filmmaker, who never married, spent his entire life addressing the same topic: family. Through his representation of marriages, funerals, and miscellaneous domestic situations, Ozu is really telling one story from beginning to end: the disintegration of the family. In so doing, Ozu actually chose the best angle from which to observe Japanese people's lives: for Asians, the family is the foundation on which social relations are built. Its gradual disintegration is Ozu's main focus in every one of his fifty-three feature films. There's no need to search far afield for one's subject; the Japanese story sits on a tatami, and so does the world's. His films tell that story unfalteringly, employing a rigorous form, with the steadiness and diligence of a well digger. The result is what Kurosawa

---

1. The Chinese calendar, adopted by neighboring countries including Japan, is lunar; it is based on the combination of ten heavenly stems and twelve earthly branches, which form a sixty-year cycle, a *jiazi*.

has called "the beauty of Japanese film" and what has become the paradigm of the so-called Asian film aesthetics espoused by Hou Hsiao-hsien and celebrated by Wim Wenders. His work is a source of invigoration for successive generations of filmmakers.

On December 12, 1903, Yasujirō Ozu was born to a middle-class family in Tokyo's Fukagawa District. His father was a merchant in the fertilizer trade. When he was ten years old, Ozu moved with his mother to his father's remote hometown, where he received his education. His father remained in Tokyo to work. During Ozu's childhood and adolescence, from the age of ten to twenty, his father was absent. Under his mother's care, he could do as he pleased. Ozu said of his mother that she was an "ideal mother." But when he reached creative maturity he also idealized the figure of the father, as we see in *Late Spring*, *Tokyo Story*, and others.

Late Spring, made in 1949, is my favorite Ozu film. It tells the story of a woman who has passed the marriageable age and lives with her aging father in Kamakura. She has no plans to marry, accepting the idea that caring for her lonely father, and depending on each other, is a fine destiny. The father, who feels responsible for his daughter being deprived of her youth and love, is considering remarrying. When she discovers that her father is planning to remarry, the daughter decides to get married as well and leaves home. After her marriage, her father, who really had no intention of remarrying, is left alone at home. In his subsequent films, Ozu adopts the same measured pace, the same use of very few characters, and minimalist mise-en-scène, which yields an abstract quality to his films. Extremely concrete and extremely abstract: that's the wonder of Ozu's films.

It is generally acknowledged that the 1953 work *Tokyo Story* best represents Ozu's cinematic aesthetics. It tells the story of an elderly married couple who live in the countryside and go to Tokyo to visit their son and daughter, both married. Too busy to host the elders properly, the children devise a plan. They send the old couple to Atami, a popular hot spring resort. On the surface, this looks like a generous filial act, but in reality, they're getting the old folks out of their hair. The only person who is good to the elderly couple is their second son's widowed daughter-in-law. When the aging mother dies, the children rush back for the funeral as if performing a duty. They hurry there and hurry back, leaving their lonely elderly father behind. When *Tokyo Story* was ranked second in *Kinema Junpo*'s list of top ten films, Ozu said, "Through the changing attitudes of the children, I attempted to portray the disintegration of the Japanese family structure." He added, "This is one of my most unsophisticated works."[2] Ozu is obviously being modest. Ozu's films are deeply rooted in quotidian life; from the opening to the closing credits, they show people's common experiences.

In 1921, after graduating from high school, Ozu worked for a year as a substitute teacher in a primary school in the countryside. He then returned to Tokyo, where he was introduced by an uncle to a manager in the Shochiku Film Company. This is how Ozu came to work for the famous Shochiku, where he remained for forty-five years. Ozu was trained as a director in the studio's traditional way, from the bottom up. He started as a gopher moving filming equipment, then a year later was promoted to assistant director, and a year after that to director. His rise can be attributed to his determination during that period but also

---

2. "Unsophisticated" is my translation of *tongsu*, which also means "vulgar," "popular," or "unrefined." Donald Richie translates the term as "melodramatic" in his book *Ozu* (University of California Press, 1974), 239. In Chinese, the term "melodrama" requires one additional word, *ju*, a "play"—*tongsu ju*—so I follow Jia's use of the term *tongsu*.

to his talent as a screenwriter and director. *Sword of Penitence*, his directorial debut in 1927, is a costume drama based on an American film he'd read about in film magazines but never seen. The negatives, prints, and script of this silent black-and-white are lost, so we have no way of knowing the specifics about this film. All we can say is that, from that film onward, Ozu took the directorial road.

Incredibly, Ozu made five films in 1928, six in 1929, and seven in 1930—a pace even more frenetic than Fassbinder's. His early films are mainly comedies, which are set against a realistic social background. They opened the way to Japanese-style realism. As early as the 1929, in *A Straightforward Boy*, Ozu had found what would become his constant theme —the family— along with the social institutions with which the family interacts: school, work unit, company, and so on. From 1930 onward, Ozu's film language became increasingly simple and unadorned: he abandoned silent film's editing techniques, such as fade-ins and fade-outs. Even while making commercial productions, he slowly created his own film method.

In 1932, Ozu made his first masterpiece, *I Was Born, But….* This ninety-one-minute-long, black-and-white silent film tells the story of eight- and nine-year-old brothers from a working-class family who discover that their adored father, an office worker, grovels at the feet of his boss. Meanwhile at school, the boss's son haunts and bullies them like the devil himself. The two go on a hunger strike but ultimately find that the boss seems like a good guy after all. This film combines all the elements of Ozu's film aesthetics and probes, through the children's point of view, Japanese society's rigid class structure. As the children lose their innocence, the film becomes more than a comedy. Ozu himself said, "I initially wanted to make a children's film but ended doing a film for adults. I wanted to do a simple, cheerful, lighthearted work, and yet, gradually it became heavier and heavier." This film was awarded first place in *Kinema Junpo*'s ten best films of the year.

In 1936, Ozu made his first sound film, *The Only Son*, and in 1958, his first color film, *Equinox Flower*. Having followed the development of film technology through a long career in the movies, this was one director who would allow himself neither to be subjugated by technology nor to be made redundant because of it. From start to finish, he worked to establish his own cinematic universe; he created a cinema to express his vision of the common people.

One of the great qualities of Ozu's films is his restraint in portraying the plight of people. Distortion and overstatement remain common flaws in the great majority of films. In contrast to his Japanese peers, Ozu resembles neither the later Keisuke Kinoshita, who is marked by romantic illusions about the family, nor Mikio Naruse, who rejects the family unequivocally. Ozu always maintains distance in his observations and restraint in his emotional response, which is far from easy. Faced with Ozu's cautiously objective position, some people claim not to like his films because they are too gentle, too bourgeois. While the Japanese may not actually be that restrained or calm in their daily lives, Ozu's aesthetic approach is nevertheless motivated by an ideal of genuine human nature. Ozu's films never go to extremes; his characters maintain their simple and authentic humanity throughout.

Through the course of his career as a director, Ozu was especially fond of shooting from one particular angle: the height of a person sitting on a tatami. Whether he was shooting interiors or exteriors, Ozu positioned his camera some three feet off the ground and usually left it stationary. Whenever the camera did move, it abruptly disrupted the immobile world he had established, creating a strange and uneasy atmosphere. Ozu's trademark shots were adopted by his Chinese disciple, Hou Hsiaohsien. These two masters' cinematographic language is rooted in their philosophy of life: immobility in observation, which in fact is a way of listening carefully and respecting the observed. Ozu also paid great attention to the film frame: before making a film, he would draw his storyboard in pencil, frame by frame. His camera was mostly positioned at right angles to the scene and at the lowest possible height. The attention to each frame and the camera position suggest that Ozu was attempting to create the effect of pictorial compositions.

With rigorous self-control in filming, Ozu created a simple, austere model. He never wavered, relentlessly repeating his theme and film technique. This was Ozu's signature. Eventually his films were recognized as anthropological works, as records of the lives of Japanese people, and as an important part of Japanese culture. Ozu's own self-restraint and his formal constraints also reflect the mindset of the majority of Asian people. That's why the beauty of Ozu's films is also the beauty of Asian cinema.

Rather than arousing emotions, Ozu's cinematic method captures them. By limiting his line of vision, he sees more; by confining his world, he transcends it. In contrast with more conventional films, the "stories" in Ozu's films are few and far between. And these quasi stories frequently seem to be about very little. Of Ozu's fifty-three films, twenty-five were made in collaboration with the famous screenwriter Kogo Noda. The eight films after *Early Spring* were conceived at Noda's country house while they took strolls, drank alcohol, and argued over the script. Ozu gives us so little of his world, yet he gives so much.

On December 12, 1963, Yasujirō Ozu succumbed to cancer. His work only gained international fame many years after his death. But fame was not something that preoccupied Ozu: his grave does not bear his name. Only a single word is carved in his stone: Nothingness (*mu*).[3] Early on, Ozu had thoroughly understood humankind.

In 1999, when I visited Japan for the first time, my producer, Shozo Ichiyama, took me to Kamakura to visit Ozu's grave at the old Engaku-ji monastery. We lit incense and placed alcohol and flowers on his grave. As I bowed, I felt the incense rise on a delicate current. I believe that Ozu's spirit was lingering in the air.

Originally published in *World Screen* (*Huanqiu yinmu*), volume 12, 2003.

---

3. This is the Japanese pronunciation; in Mandarin Chinese, it's pronounced "*wu*."

WE NEED TO RECOGNIZE THE DEFECT IN OUR GENES *(A LECTURE)*

25

Translated by Alice Shih

Greetings. I am delighted to be back here at the school. This year marks "A Century of Chinese Cinema," and I have been participating in many events. Each time, I'm confronted with complex emotions. One hundred years absolutely calls for a commemoration, which everyone who loves Chinese cinema should be excited about. Furthermore, we might have been a bit neglectful of our early cinema. Most films produced in the 1920s and '30s have been confined to vaults. We may have some DVDs of those films, but they are still sitting inside our cabinets.

I learned film theory at the Beijing Film Academy in 1993, and to this day, I still can't tell you how much I fully understand about the past hundred years of Chinese cinema. From the silent period to the modern day, there are many high points in the history of Chinese cinema that remain unfamiliar.

We put together a commemorative event for this occasion, like an impious son who has long neglected his parents but then mounts a grand opera to celebrate his father's sixtieth birthday so that everyone sees him as a loving son. He then reverts to his old ways. We should not be like this. We should think of this day as a new beginning. We should study and treasure our precious film tradition. We should analyze our films through frequent screenings and intellectual discussions.

It doesn't matter if a film is old or newly produced. It is new as long as audiences haven't seen it. A film made in the twenties or thirties may have existed for eighty years, but if you haven't seen it yet, it's a revelation. That is why we all gather here today for this grand event, so we can spend an afternoon discussing this important matter.

A hundred years of cinema can open our eyes to many cinematic possibilities. As a director, I can't just look to contemporary cinema for inspiration. The spirit and technique used in the pioneering silent films of the twenties are equally enriching.

Filmmaking was in its childhood back then. Production technology was not well developed and was begging for new ideas. Every pioneering director was searching for a cinematic language without recourse to tradition or experience. In fact, all the questions that have arisen recently in Chinese cinema, all our recurring dilemmas, could be solved or informed by our traditional cinema. I acknowledge that learning from Western and contemporary cinema is essential, but understanding our own traditions is just as important.

Looking back at my journey to understanding filmmaking, director Yuan Muzhi's *Street Angel* (*Malu tianshi*) (1937) set a certain standard. Through it, I understood the mistakes made in Chinese films after 1949 and before 1989. The greatest mistake was the domination and mainstreaming of ideological revolutionary arts. The rapid progress of film production in the thirties and forties was halted, and *Street Angel* was put on hold for the inspection list. The film was also subsequently misinterpreted. We were told, in bureaucratic parlance, that it is a "leftist" film. Of course, *Street Angel* is a leftist film, but we can't deny its high artistic merits—merits that would not be found in films produced after 1949, merits that were gradually eliminated, neutralized, and forgotten. One of the artistic virtues of our cinema tradition was the savvy and vibrant depiction of street life. The

WE NEED TO RECOGNIZE THE DEFECT IN OUR GENES (A LECTURE)

creative direction of the film unfolded through subtle character development. Zhao Dan's character of a trumpeter, and Zhou Xuan, the street singer girl, establish an important human relationship. The story is centered on this human relationship, and events take place in familiar quotidian places; the film pays homage to the dignity of real human experience.

We often speak of films made by Yasujirō Ozu and Hou Hsiao-hsien as paradigms of this tradition. In fact, *Street Angel* paved the way for them. Yuan's genius was to write and direct a film that depicts life within a lively marketplace, presenting it through skillful mise-en-scène. The film opens with documentary footage of the Bund in Shanghai. After the crowd exits, a band enters the frame and Zhao Dan, the trumpeter, walks into the crowd. Yuan skillfully highlights the spatial relationship between the streets and all the neighbors in the marketplace, one by one as they are introduced. We travel from the newspaper seller on the street to the singer girl played by Zhou Xuan, then to her standing on the second floor greeting Zhao Dan. The camera then turns to take a master shot of the tea house, her workplace. This opening sequence looks simple, but it captures the familiar rhythms of people attending teahouses, constructing an energetic mise-en-scène that details the daily activities of ordinary people.

This cinema tradition was broken after 1949 when our individual rights and leisures were replaced by communal living. Chinese communities were no longer tied to streets or locations. People were grouped together in compounds and units, and we were defined by this new community structure. Much of the population had to leave their familiar streets, teahouses, and neighborhoods behind.

During and after the ten years of the Cultural Revolution, a few people were able to stay in their old communities, among their neighbors. However, the centralized, homogenized communal system left individuals with little privacy, and the old idea of open relationships on the streets with neighbors, as depicted in the film, was considered frivolous and negative and eventually disappeared. Adopting such a way of life, we even neglected our own rights and richness of experience. Losing touch with stories structured around human relationships, Chinese films turn toward the genres of melodrama and fantasy. Since melodramas appeal to the masses, this combination of genres was used to satisfy the popular market. Most films rely on legendary tales or black-and-white cinematography to increase the sense of drama, as well as clichéd plots like "till death do us part."

Some important works were produced after 1949, including *Guerrillas on the Plain* (*Pingyuan youjidui*) (1955) and *Railroad Guerrilla* (*Tiedao youjidui*) (1956). They are a combination of revolutionary art and legends. They are extremely dramatic and generally admired, and they mark the birth of a new film genre. Our cinema no longer deals with individuals under the pressures of daily life. We have become ill equipped to deal with this kind of subject matter.

This stifling of our cinematic tradition has presented challenges to many young directors trying to tell personal stories. Some might even think that certain lives or locations should not be portrayed on screen at all. They fail to recognize their value or how to translate them to the screen. They say, "Is it justified to spend so much money on a film just talking about personal feelings?" I think differently: there isn't any particular emotion that should be off-limits in the creative arts, regardless of how personal or private it is.

When creative works about personal emotions and even bodily sensations are phased out, they are often replaced by ideology. Many brilliant films, including classics from 1989, are the products of ideology, works of group thought. Works from a personal point of view, which look into the society or history of China, whether in the form of a lament, a celebration, or a scream—all these expressions have become vague and unfamiliar. This complicated cultural experience is a part of our post-1949 reality. We live in this vast country steeped in important historical events, teeming with cultural relics, literature, and institutions. We often take our cultural resources for granted. We assume they are merely inherited, outdated orthodox traditions. Many people, including me, are skeptical of this passed-down knowledge. These traditions are widespread in the Chinese diaspora, and many of them are better preserved in Hong Kong and Taiwan.

Today, we enter into a new era. We should not be examining films alone. We should start watching films to learn about Chinese culture in general, to get a more complete picture. We should break through our conservative cultural background and illusions and open our minds to analyze and learn this culture from all angles.

A few mainland films show a sensibility similar to *Street Angel*. One is *Swan Song (Juexiang)* (1985). Another is *Good Morning, Beijing (Beijing ni zao)* (1990), which was made in the eighties and early nineties. But things were different in Hong Kong. Their commercial cinema abounds with films portraying dignified and complex human relationships. In *Endless Love (San bat liu ching)* (1993) you can see a creative skill and flexibility that Hong Kong filmmakers have, which we lack. One reason may be that Hong Kong cinema maintained a link to the Chinese cinema tradition of the thirties and forties. If we look into film history, we can trace a lineage that started when the filmmakers in Shanghai fled south to Hong Kong at the dawn of WWII. There it survived past

1949 to this day. The Hong King film tradition also preserved and developed martial arts and swordplay (*wuxia*).

We Mainland Chinese filmmakers, who work within this atmosphere and context, need to do more research. We need to appreciate these precious films made in other Chinese regions. We must appreciate cultural traditions unfamiliar to us. In *In the Mood for Love* (*Fa yeung nin wa*) (2000), director Wong Kar-wai unveils his personal film philosophy and shows us his reflections on Chinese culture. He was born and raised in Shanghai and understands Chinese traditions. In particular, he features the cheongsam (a traditional Chinese dress). The film ushered in a renaissance of that fashion, but it didn't spring up from nowhere. It was a result of the director's cultural inheritance. It is hard to imagine that a director who had gone through revolutionary education and cultural renunciation would have the passion to make this kind of film.

Human emotions are raw. Some film concepts are more rarefied. For example, we have the term "big air" (*daqi*).[1] Why is "big air" a good aesthetic standard? Or the term "powerful." Why is it positive to be powerful? We tend to judge all art according to how much "big air" or "power" it conveys. Thus, softness and private, gray, personal matters are considered negative traits or artistic mistakes. When shooting a film, or in any creative endeavor, I believe that all kinds of artistic temperaments are needed. Perhaps all Chinese filmmakers, especially those in the Mainland, should study the past century to diagnose the flaws in our film tradition. These flaws are worth more of our attention and contemplation.

Century of Chinese Cinema lecture: Originally published in the *Beijing News* (*Xin jing bao*), December 17, 2005.

---

1. "Big air" (*daqi*) is a Chinese slang term describing a display of regal, masterly, effortless skill.

MARTIN SCORSESE, MY "ELDER" 26

Translated by Alice Shih

It was 1996 when I first went to Hong Kong. I was still a student at the Beijing Film Academy. I had no idea that foreign directors' and actors' names were transliterated completely differently in Mainland China and Hong Kong. I had no idea who they were referring to when the names of Godard or Truffaut were mentioned in Cantonese. I had just made the acquaintance of Yu Lik-wai at the time, and we would hang out at teahouses and restaurants discussing arts for hours. We spoke of our favorite directors, and he went on telling me about this director named "Ma Tin" for half an hour. His description was quite vague and he didn't specify what films he had made, so I asked him directly who he was talking about. He said he did *Taxi Driver* (1976) so it dawned on me that he was talking about Martin Scorsese. In Mandarin we call him "Ma Ding." I thought he had been talking about some Chinese director with the last name of Ma. We laughed out loud and realized that we were lost in translation.

When I was still learning the craft of filmmaking at the Beijing Film Academy, Martin Scorsese's *The King of Comedy* (1982) had given me a lot of inspiration. I inferred it was probably shot on multiple cameras, and I realized it was possible to edit together different camera angles. Usually, I like to use mise-en-scène to highlight spatial relationships, to visualize and make use of the actors' stage blocking to illustrate the narrative. *The*

King of Comedy was different. He filmed the same action from multiple angles and edited these different angles together to create an interesting visual effect.

Information on filmmaking was hard to find in 1996. We had a lot of respect for the masters, but to us they were all unapproachable and mysterious. To us Chinese film students at the time, Martin Scorsese was like a shining star, transcendent and inaccessible.

I brought *Unknown Pleasures* to the Cannes Film Festival in 2002, and Martin Scorsese was the head of the jury for the shorts program. When I heard that he was around, I was thrilled and was hoping to run into him.

One evening at a festival dinner, I heard a big commotion outside, and then a petite older man entered the room. He attracted a big crowd, and his entrance was like that of a governor. I didn't recognize him as he weaved through the crowd to search for his seat. My Japanese producer elbowed me and said, "That's Martin Scorsese!" All of a sudden, Scorsese was next to us and I stood up to shake his hand and I told him that I am Chinese. By then, people had flocked to him from all directions. I observed that he held eye contact with everyone who came to him, and he took time to respond. When I was in front of him, he said "China" and "Great Wall Hotel." I repeated "Great Wall Hotel" in English. He explained that he was in Beijing in the eighties, and he stayed at the "Great Wall Hotel." That was our first encounter. After that, he was swamped and was taken away for other matters.

One day after I came back from Cannes, I was awakened by a fax. I picked it up and was surprised to find that it was from Martin Scorsese. It was written, "We met at Cannes, but our meeting was cut short. My assistant told me afterward that you are the director of *Xiao Wu*. I saw *Xiao Wu* and I liked it very much. I didn't realize who you were at the time and didn't get a chance to chat with you. If you happen to be in New York, please call the number below and we can meet again."

Our chance came quickly in August when I received an invitation to bring *Unknown Pleasures* to the New York Film Festival. I had never been to New York and was eager to go. My producer, Chow Keung, graduated from school in New York almost ten years earlier and hadn't been back. So, Chow and I decided to take the trip together. Both of us were excited to have a meeting with Scorsese on our itinerary. I sent Scorsese a fax telling him when we would be in New York to present *Unknown Pleasures*. He replied promptly saying that he was busy editing *Gangs of New York* (2002), but he would find the time to meet with us. In October, we packed our bags and boarded the plane to New York.

The hub of the New York Film Festival is Lincoln Center. Ken Jones, who was on the programming team, came to find us. He has written extensively about Asian cinema and is a very important figure in presenting Asian films to America. I got to know him in 1999 at the San Francisco International Film Festival when he was a member of the jury. He honored *Xiao Wu* with the Best Picture award and handed Martin Scorsese a video copy of the film, as he was also Scorsese's assistant. That's how Scorsese got the chance to see my film. Since then, Scorsese has introduced *Xiao Wu* to different people, including Steven Spielberg, and has spoken about it at college lectures. It is interesting to me that there doesn't seem to be a barrier between commercial films and art films in America—Spielberg and Scorsese both came from an experimental film background.

We arrived at Scorsese's office building and entered a crowded elevator without a word. His tall assistant, Ken Jones, was showing us the way. I looked up to see him chuckling at me, and I felt a bit uneasy. I turned around only to see Martin Scorsese right next to me. We broke into laughter and my anxiety dissipated instantly.

We entered Scorsese's office to find a table of snacks. It was like visiting an elder when I was a little boy. We sat down and Chinese tea was prepared. He had bought the biscotti particularly for us and urged us to try some. That was when I really felt like I was visiting an elder and not meeting about films. He kept smiling at us, like he was in the company of children. He did not eat anything, but we ended up eating a lot.

We started our conversation on *Xiao Wu*. I thought he would start from a professional angle, but he did not. He asked if I knew why he liked the film. I thought that maybe the social stratum and situation of my characters were similar to his early works. He said that I was half right. He had an uncle who really resembled my character. This uncle worked in Little Italy and Scorsese once went to him asking for an opportunity to make some money. He was told to come back the next day, and when he did, he was given an address and a gun. His uncle wanted him to collect a bad debt. He told me he was trembling with the gun in his hand. Luckily he did not turn into a debt collector, or America would have lost a film director to the mafia. He thought Wang Hongwei's portrayal of Xiao Wu in the film was very close to his uncle and found him very endearing.

When you have become an established director, it can be difficult to revert back to being an ordinary member of the audience. But that did not happen to Martin Scorsese. When he discusses filmmaking with you, he doesn't comment on complicated mise-en-scène techniques or narrative structures but about his uncle and his past. He could be totally immersed in a film, but he could also come right out of it.

MARTIN SCORSESE, MY "ELDER"

During our conversation, a courier came in and delivered a package. It was an assignment from a film student at a university in California. I asked if he was teaching there and he said no. I wondered why students were sending him work, and he explained that his office accepts unsolicited works, anyone can submit. I was deeply moved. I remembered needing a lot of help when I first started making shorts. He insisted that it wasn't he who was helping the students; he couldn't really help them much. It was, in fact, the students who helped him. These works made him realize that he was a throwback from the last century. The student films gave him fresh ideas and helped him keep up with the times. After he said this we felt moved and were quiet.

After that moment of silence, he said he had to get ready for more editing and he asked Ken Jones to take us to his film vault. He has hundreds of 35 mm and 16 mm reels, LD, VCD, DVD, and VHS. The titles span decades. I saw works of Antonioni and asked Ken Jones if he really watches them. Jones said he does frequently. We then toured his screening room where his assistant said he spends a lot of time watching his favorite films. I had decided right then that I would print a copy of *Xiao Wu* for him as a gift.

Scorsese's work environment and office layout give the impression that he is living in a world of films. This space operates not only as an office but also as a production workshop. It is a toy room for a mischievous, passionate film lover. After the tour, we saw that Scorsese was ready for his edit session. Working next to him was an elderly lady, and he introduced her as his longtime editing partner since the seventies when he was still making shorts. A man who looked like a taxi driver then entered the room, and Scorsese introduced him as his longtime screenwriting partner who he had been collaborating with since *The Age of Innocence* (1993). He has his own film crew, and they are all old friends and partners. Their relationship is not that of employer and employees but artistic collaborators who come together to create art. At the same time, they are like family with emotional ties.

I watched him work in the editing room; he was working on the opening action sequence. He suddenly stopped and said to me, "I watched *Battleship Potemkin* yesterday, and I have lots of questions for Eisenstein." He then stopped and tuned back into his footage. He has knowledge of the entire history of cinema. He doesn't just know about contemporary films, he knows all about the silent classics of the 1920s. His film references are informed by the whole history of film watching. I observed closely how he talked to his team; they communicated tacitly. After nearly an hour, I felt like I had taken up too much of his time and decided to leave. He walked me to the door and before I stepped into the elevator, he suddenly gestured to me. I approached him and he told me, "Keep your budget low." Then he turned around and stepped back inside.

On our way down, no one spoke inside the elevator. Everyone was figuring out what he meant. For him, keeping a low budget upholds his ideal. It helps a filmmaker maintain his freedom and creative control. Budgets were something he had to deal with at the time, and he emphasized the advantage of keeping it low. *Gangs of New York* had a budget of ten million. The investors wanted him to shorten the running time and they had some conflicts. His line to me was an encouragement as well as an expression of his own creative frustration. I felt sympathy in my heart for this great veteran.

Originally published in *LIFE Magazine*, volume 3, 2006.

EVERYONE'S BODY IS A WORK OF ART IN ITSELF

27

A Conversation between Liu Xiaodong[1] and Jia Zhangke
Conducted by Deng Xin and Wang Nan
Translated by Alice Shih

THE AESTHETICS OF HUMANITY

Deng Xin: Master Liu, you have been to the Three Gorges area a few times. Where did you go this time? Did you feel differently on this trip?

Liu Xiaodong: We went to Fengjie this time. The whole town is about to be completely immersed. I couldn't even recognize the places I visited a couple of years ago.

---

1. Liu Xiaodong, a renowned contemporary painter, is well known for his meticulousness and his realistic depictions of people in modern living conditions. He was invited to numerous important exhibitions in China and abroad in the late 1980s. His work has been exhibited around the world, including the National Art Museum of China, the Museum of Contemporary Art Shanghai, the San Francisco Museum of Modern Art, the Fukuoka Art Museum, and the Queensland Art Gallery.
  At the end of 2002, Liu Xiaodong traveled to the Three Gorges region. He saw the progressive demolition of the region and was inspired to paint it. In 2003, he painted the first part of the series, *Great Immigration at the Three Gorges* (*Sanxia da yi min*). In 2004 he added *The New Immigrants of the Three Gorges* (*Sanxia xin yi min*). Late in 2005, Liu Xiaodong and Jia Zhangke went to the Three Gorges together. While there, Liu painted *Warm Bed* (*Wenchuang*), and Jia turned Liu into the subject of his documentary, *Dong*. Jia also got the idea for another film, *Still Life* (*Sanxia Haoren*).

Jia Zhangke: I suppose only about a quarter of Fengjie is still remaining.

Deng: The Three Gorges Dam construction started a few years ago. How are the lives of the people who were affected by these changes?

Liu: There aren't many changes for them. They are still living the same way in the same conditions. It doesn't matter where the common people of China are—whether Tibet, Kashi, or the little town of Fengjie—people everywhere are the same. But compared to our living standards, they go through a lot of hardships.

Wang Nan: But what I see in your paintings are exquisite people of peace and warmth. At times I could even catch a trace of satisfaction in their spirit.

Liu: I really like to look for joy in difficult situations. I search for happy people or things to paint, not misery. Somehow, every cloud has a silver lining.

Deng: That seems to be different from Master Jia's films. They always carry a bitter taste. What brought the two of you together for this collaboration?

Jia: It wasn't because we wanted to do a project on the Three Gorges. It was because Liu Xiaodong had to go there to paint, so he invited me to shoot my documentary about him. At first, I was just curious about his work conditions and routine and the way he interacts with his models. I had never been to the Three Gorges area, but I fell in love with the region the first time I saw it in Liu Xiaodong's paintings. The place intrigued me, so we decided to go there together.

Deng: Can you describe behind the scenes of the shoot?

Jia: We shot his painting process, the way he expressed his feelings about the region. Liu Xiaodong has very good reasons to choose this region. It is disappearing and changing. Who we see today might not be around the next day. Maybe he died or moved away. Liu had to keep up with these changes, like chasing the traveling sunlight. The place he was painting had a tall building behind it, so if he didn't hurry, the light would be blocked and he wouldn't have enough light to paint. I was drawn into his painter's world.

Deng: No doubt you have gone further in understanding his art and soul through this collaboration. Can you tell us more about it?

Jia: What moved me was not his choice to paint the Three Gorges area. I was inspired by the way he treats life itself and his love for the individual people. The people Liu Xiaodong worked with were all demolition workers. He captured a particular kind of beauty in their bodies on his canvas. His heart is full of empathy for the people he treats with his brush. This is what moved me the most on the trip to the Gorges. I had shot documentaries before, but I haven't fallen in love like that. He had an effect on my mind while we worked closely together, so I decided to make another narrative feature that was completely unplanned and it turned out quite differently from my previous films. While watching him paint, I got the feeling that it doesn't matter what kind of environment we're in, the body given to us by the creator is ultimately beautiful.

I consider myself an art lover and I have always wanted to uncover the secret of his painting. Now, when I look back at his paintings, they all seem to have a shining quality; each person he paints possesses a unique beauty. Jargon such as "realistic approach" or "realism" does not do justice to his works; they are a direct expression of love toward their subjects and to life. They are very natural and raw. It is hard to find something so genuine in our over-packaged world. We seldom encounter such character. The title of this documentary is *Dong*, meaning "East," which is where we are situated in the world, and it happens to be his name.

Deng: How did you design the structure of *Dong*?

Jia: The film has an open structure loosely divided into three sections covering his painting locations, thoughts, and interactions.

Deng: Why would you choose to paint in the Three Gorges area?

Liu: I have always been afraid of my own vanity, like when I had achieved something and I became arrogant. I have often tried to make myself more humble, think less of myself, and be less opinionated. An artist has to make a connection to the livelihood of the local community before he can create fresh and powerful works. If I didn't go to the Three Gorges, I would have gone to some place similar. In Beijing, people are paying attention to me as a painter, and I felt like I was contributing to society through things like exhibitions, interviews, and book launches. But when I landed in the Three Gorges, I faced a whole town about to be submerged. In that situation you feel like you are totally helpless, and vanity becomes meaningless instantly. You re-evaluate many questions. When artworks are worshipped, an artist feels empty. In the Three Gorges area, artworks are not worth the price of a mattress. Nothing is more useful than a mattress. We couldn't change anything, and we had no impact on their lives. I was hoping to treat myself to a humbling experience, not to take myself too seriously.

Wang: Is *Warm Bed* (*Wenchuang*) the only painting you did at the Three Gorges?

Liu: Yes, it measures 10 m. x 2.6 m. [32 ft. x 8 ft.]. It shows eleven shirtless peasants in shorts sitting on a big mat, playing cards against the mountainous background of the Yangtze River.

Deng: *Warm Bed* is a huge painting. How did you control the crowd and location?

Liu: I divided the painting into five panes. I did not do a preliminary sketch but worked on it pane by pane. Doing large-scale work is more interesting for me. For smaller works, you can record everything in an instant, but this is impossible for big paintings. After you finish drawing the people, the scenery will have changed already. It was like racing with time when I drew it. I consider sketching live on canvas the happiest thing about painting.

Deng: The combination of the documentary, *Dong*, and the painting, *Warm Bed*, could be considered one piece of art. Like the big painting of *Great Immigration at the Three Gorges* (*Sanxia da yi min*), it might be hard to capture a scene so grand and a theme so complex on a single canvas.

Liu: There's always a limitation in the size of a canvas, which is why I returned to the Three Gorges area a few times. It wasn't only to document and record the situation but also to express my concerns and my feelings about the people there. Of course there was more power there than this piece could express.

FEELINGS TOWARD THE THREE GORGES

Deng: Can you tell us more specifically how the Three Gorges made you feel?

Jia: Our work came together naturally, both artistically and emotionally. Most important, we got more intimate with humanity. We got to understand why we like to reach out to people and what was the best approach to express the living conditions of these people, who are relatively poor. They all rely on this river to survive, and their livelihood will be taken away after the relocation. Yet, when we came to specific individuals, we discovered that happiness could not be measured entirely by wealth. We came across some people who have great contentment in life, who were very easy going, without much worry, anxiety, or sorrow. They live happily, regardless of how much they earn for the day, and that's a blessing.

After we came back, many in our crew found themselves missing the Three Gorges area. They found their old ways of living, with all the drinking parties in the city, not as fun as before. They felt an emptiness and pretense in Beijing. Materialistic gratification failed to satisfy them after experiencing the feeling of simple living in the Three Gorges area. They seemed to have found the source of happiness, which is very important. I'm not too concerned about the distribution and audience reach of this film in the future, because I was so happy during the course of making it. It was like having a child: you really treasure this new life, and you are delighted just watching this kid grow. The most cherished thing is the love that brought this life into the world and the unforgettable process of birth. I was more confident, grounded, after I came back, and I appreciated my life more.

Liu: I share this feeling. In the Three Gorges area, individuals do not have complete control over their own lives.

I was able to express my internal feelings through the landscape and the people while I was there, using my paintbrush on the canvas to reflect their mentality and destiny.

ON FILMMAKING

Wang: You had a cameo appearance in Jia's film *The World*, and you are familiar with the filmmaking process. It looks like you really take an interest in films.

Liu: I think people working in film and television are interesting. They have a lot of innovative and vivid ideas, so I like to be with them. Filmmaking is, in fact, extremely difficult. Jia would shoot from 5 a.m. to 3 a.m. the next morning. It seems to me he really enjoys that profession, whereas I would have collapsed. He has to make a lot of decisions aside from purely artistic ones.

Jia: Filmmaking is a systematic process; it needs the cooperation of a team and is impossible to do on your own. I need to collaborate with many people and we make all kinds of decisions. A painter seems to enjoy more freedom. You have more control over time and the way to do it.

Deng: Your work shares a common thread: the portrayal of the concrete lives of the general public.

Jia: The people in my films are people you would run into on the streets—very similar to the subjects in Liu Xiaodong's paintings. They all look ordinary, but they possess their own beauty, the kind that they were born with, and that quality can be very moving.

Deng: Director Jia, I once read an interview where you said, "I'm against films that look perfect, like a painting, where the highly polished form overtakes the aesthetic content of the film. When it comes to spatial design on screen, I want to emphasize the portrayal of subjects and the atmosphere around them. It shouldn't resemble a well-composed painting." I have some training in painting, and I find it hard to understand. Can you elaborate on that?

Jia: That remark was made at a time when Chinese films were being judged according to how well they achieved a visual effect that resembled paintings. I find that absurd. It's not that I don't like paintings, but on-screen treatment is totally different from what you do on canvas. A series of frames are connected in films, there is a narrative chronology. Their resemblance to paintings should never be a standard of judgment for films. Even though a film might possess certain aesthetic qualities shared by paintings, that is not all there is to the film. Analogously, we can't find characters in Chinese movies that resemble the people in Lucian Freud's paintings. Most people wouldn't consider these faces "beautiful," but Freud is universally recognized as a master painter of these "not beautiful" portraits.

Deng: Director Zhang Yimou's recent works are saturated with vibrant colors to heighten the visual aesthetics, but that has generated mixed reviews. What is your opinion on that?

Jia: It is hard to compare apples and oranges, but every individual film must be judged by its achievements. Good works of art will be accepted and praised. I agree that we need our own film industry now. I have a lot of respect for directors who excel in commercial filmmaking. My films are more than that; they carry my habits as well. I don't live life in the fast lane that much, so I take my time with my shots. I think films should be more observational and filled with colorful vignettes of the people in the world.

There should be a bit of local flavor, even in commercial films. In fact, martial arts films are inherently Chinese, but it seems like this characteristic has been neglected lately. Protagonists in martial arts films may or may not be Chinese now, and it's all mixed up. I think some directors should do something about it. "Crowd-pleasing" is a seductive, dangerous trap. Many consider that when a lot of people are doing it, it should be acceptable. But it may not be.

Deng: Can you say that your films are purely a reflection of your feelings?

Jia: I couldn't say they are *that* personal, but they do carry a lot of my thoughts and style. I don't see the essence of life being so complicated, so why make films so dense?

Film may be the only medium of expression where the process traverses a complex series of conditions. That is meaningful itself.

Filmmaking is actually very simple: the apple doesn't fall far from the tree, meaning the film should really reflect what the director observes. It's just a way to observe the world.

ON HAPPINESS

Wang: What do you find most attractive in Liu's paintings?

Jia: Liu Xiaodong's painting style has been changing in the past years, but whatever the changes are, he's always able to find the most meaningful aspect of his subjects, their most spiritual essence. Not everyone has the ability to seize that. If he didn't change with time, his point of view wouldn't change, or his positions and interests. Yet, he always finds new things to display through the years, always grasping people's desire to live. Say there was a painting of someone kicking a ball around as his morning exercise. For us it's an everyday sight, but simple daily routines in his eyes become poetic celebrations.

Deng: Painters are all poetic, right?

Liu: I believe that painters should be poetic, as I adore poets myself. I regard poetry as the highest form of art. By "poet" I don't mean they all have to write poetry, but they should be able to think freely and not be limited or defined by others.

Jia: In fact, everyone could be poetic. You can paint, sing, or dance to express yourself. It's hard to imagine artwork without being poetic. When you encounter a poetic moment like the setting of the sun, it should stir up an indescribable feeling in you. This is a form of poetry that anyone can experience.

I tried my hand at poetry and painting before, but I failed horribly so I turn to filmmaking. Everyone has his or her strong suit, and it's really not up to the individual to decide what art form best suits that person. It depends on the talent inside. I was crazy about poetry in my high school days and I wrote three volumes of poetry. Then I tried writing novels and filmmaking. I think filmmaking suits me best.

Wang: You seem to be able to find a moment's spiritual essence, which you analyze and capture in your paintings. What is touching and appealing to you these days?

Liu: I'm still being touched by lots of emotions. I find myself enjoying being lazy, and my thoughts wander these days. Maybe I overexerted myself previously. I take breaks between brushstrokes and naps when I'm tired. I calm myself this way. It's like calligraphers who need to write to calm their nerves.

Wang: What do you consider true happiness?

Liu: As long as I'm not just wasting time staying put, then I consider staying put a state of happiness these days. Watching television dramas at home is a form of happiness. There are a lot of nagging chores lately; just doing a couple in a day turns my world upside down. I hate that paintings are becoming a kind of luxury good, but there is little I can do about that. Come to think of it, the whole idea behind paintings is actually antimaterialism, yet I'm enjoying the material gain as I paint. This hypocrisy makes me anxious.

Looking back, it seems like I was happier before, even though I was poor. The rice cooked using a cheap electric stove smells good and satisfying. Now my conditions have improved, but that comes with troubles. Attention to material needs is actually quite a horrible thing. Something happened today: I bought two antique chairs and sent them to Shanghai for repair. I had them sent back after the work was done, but when I opened the box, I found them broken. I was very angry. Later, I was enlightened by this experience: I should not be fixed on acquiring exquisite things. When that exquisite lifestyle is challenged, we may not be able to bear it psychologically. Failed pursuits hurt when they slip out of your control. Painting can liberate me from troubles. A few brushstrokes in the afternoon can erase unpleasant memories. I did want to chase the high life, but I got more troubles in return. Happiness can't be bought.

Originally published in *Oriental Art* (*Dongfang Yishu*), volume 3, 2006.

RECOGNIZING THE BEAUTY OF ONE'S OWN BODY   28

A Conversation between Tony Rayns[1] and Jia Zhangke
Translated by Alice Shih

Rayns: Why did you accept Liu Xiaodong's invitation to visit the Three Gorges area? Was that your first visit? What did you think about it when you got there?

Jia: I had never been to the Three Gorges area before the shoot. All I knew of the area I learned through the media. People were debating if the dam needed to be built, and then there were discussions about location and environmental and cultural preservation issues. Yet, construction marched ahead. Within two years, towns and cities with more than two thousand years of history were knocked down one by one. After these cities went under water, millions of people were forced to relocate, and the area instantly went quiet.

After the plan became a reality, the previously passionate media lost interest in the region. Those quiet locals were left alone to deal with consequences of the huge construction project. When I brought my camera to the area, the demolition of the old towns along

---

1. Tony Rayns is an English film critic, festival programmer, and author. He has researched and written books on directors such as Seijun Suzuki, Wong Kar-wai, and Rainer Werner Fassbinder.

the Yangtze River was gradually coming to an end. However, the new housing project up on the hills was not completed yet. Looking at the demolished ruins in the land about to be flooded contrasted with the newly constructed tall buildings up above, I wondered if it was all a catastrophe or the dawn of a new beginning. I didn't hesitate to agree when Liu invited me to go with him, as I really love his work. Having the chance to be in the same place as him and to work side by side with different media to make art on the same theme was very exciting to me.

Rayns: You shot two films at the same time, one narrative and one documentary. What is the association between the two? Do you want these two films to open together?

Jia: I love the suggestion of the audience watching *Dong* and *Still Life* together. Our initial plan was to make just the documentary, *Dong*, but during the course of our shoot, my head was filled with vivid images of the locals. They take initiative in their daily lives, like they call job hunting "striving to survive." They recognize that life is difficult, which fuels a kind of vitality. The difficulty of survival does not conceal the beauty of life itself. When I looked through the viewfinder during the shooting of *Dong*, I saw the people inside the lens rushing back and forth in their daily lives. I began to imagine the real lives of these quiet people after they stepped away from the frame and the stories about their hardships that they did not really talk about. That's when I started shooting *Still Life*. Locations that appeared in the documentary also appear in the narrative film, and many of the characters are the same.

Rayns: Liu Xiaodong painted two sets of paintings, one of the men in Sichuan and the other of the women in Bangkok. There are also two parts in your film *Still Life*, the first half of a man and the second half of a woman. Is this a coincidence?

Jia: This is purely a coincidence. Maybe we are both influenced by Chinese philosophy. In Chinese culture, the world is made up of Yin and Yang, with Yin representing female and Yang male. This concept contributes to the fundamentals of Eastern aesthetics. While shooting these two films, I was interested in the way people experience certain physiological states. The sweaty bodies of the construction men and the singing women in the humid air of Bangkok made me appreciate the aesthetics of the human body. My previous films are more concerned with the way society affects the characters. I didn't let the two stories in *Still Life* intersect because I felt that in the past, people interacted more easily, but now people are increasingly withdrawn and lonely, with fewer chances to cross paths with others.

Rayns: Do you keep up with contemporary Chinese visual arts? Do you see an overlap in the interests of visual artists and filmmakers?

Jia: I had always hoped that films could one day become a part of contemporary Chinese arts—that they would have the same creative freshness. With digital technologies entering China, some Chinese film producers are able to break away from the limitations of the film industry. These films enjoy a greater degree of creative freedom, and the line between visual arts and traditional filmmaking becomes blurred. We are even seeing some hybrids. I think we need some rebels in every art field, especially in today's China where commercial culture is dominant. This is my fourth digital film.

Rayns: Liu Xiaodong said that your films are partly driven by sexual desire. Do you agree?

Jia: Perhaps my films are sex films without sexual images.

Rayns: Do you see an important differentiation between shooting narrative films and documentaries?

Jia: For me, narratives can better illustrate the reality in human relationships, because documentary subjects tend to deliberately hide from reality. On the other hand, in documentaries I can pay attention to the state of my subjects' bodies—the way they walk or a noise that breaks the silence of a lonely landscape. I like how documentaries help me discover and express the abstract part of life. Of course, there are other positive things about documentaries—the freedom of the shooting makes me appreciate the fundamental charm of film.

ON *USELESS*   29

A Conversation between Tony Rayns and Jia Zhangke[1]

Tony Rayns: What were the origins of this documentary, *Useless*? How does it relate to your last documentary, *Dong*?

Jia Zhangke: *Dong* was the first film in my "Trilogy of Artists," and this is the second. *Dong* was about the painter Liu Xiaodong, and this is about the fashion designer Ma Ke. After the events in Tiananmen Square in 1989, Chinese intellectuals found themselves once again marginalized. The general public lost interest in what intellectuals were thinking and saying, and only a very small, closed community gave serious thought to the future direction of Chinese society. The public at large plunged into a new era of consumerism. In the midst of all this, the Chinese contemporary art scene miraculously retained its vigor, and many artists kept thinking long and hard about Chinese society. Their insights are obviously interesting, and I've long wanted to make films to introduce them and their ideas about society to a larger audience. Hence my "Trilogy of Artists."

I first became aware of Ma Ke when she was in her studio in Zhuhai, preparing to show her Useless collection at the autumn/winter *Paris Fashion Week*.

---

1. Tony Rayns kindly gave us his English version of the interview.

Her work went far beyond the image I had of fashion design; to my surprise, I found that her Useless collection made me reflect on China's social realities, not to mention history, memory, consumerism, interpersonal relationships, and the rise and fall of industrial production. At the same time, the idea of making her into the subject of a film gave me the chance to look at a wide range of social levels as I followed the process from design to manufacture to exhibition in the garment industry.

Nobody commissioned this film, and I made it in conditions of complete freedom. Making it was an entirely pleasurable experience, thanks to my interaction with the artist.

It was the same with Liao Xiaodong on *Dong*. In both cases, the artist-subject and I faced the same reality and started from the same point. They worked with paint and fabric and I worked with film.

Rayns: Your early fiction films were all set in your native Shanxi Province, but more recently you've transplanted Shanxi characters to other parts of China and you've looked for broader perspectives. *Useless* is a bit like *The World* in that it sets Chinese realities in a global perspective. Is that how you see it yourself?

Jia: Since making *The World*, I more and more like to use block structures and to represent more than one group of characters or more than one setting. *Dong*, for example, brings together two Asian cities that are very far apart, Fenyang in the Three Gorges area of

Sichuan and Bangkok in Thailand. And in *Still Life* I told two unrelated stories that happened to take place in the same town. As I grow older I experience life's complexity and diversity, and it seems to me hard to represent those characteristics through a conventional linear narrative, such as a ninety-minute story of one man and woman in a relationship.

These days it's no longer unusual to travel, to meet a wide variety of people and to experience very different kinds of relationships. Our sense of the world changes as we travel and cross-reference different realities and lives. Obviously, low-cost air travel, satellite TV, and the Internet all contribute to this changing sense of the world. In many parts of China, it's already the case that most people no longer know only the immediate reality around them. In the film, I show that Ma Ke created her Useless label to protect against the industrialization of garment making on a mass scale. In Shanxi, garment workshops in remote mining areas are dying out because they cannot compete with the vast garment factories in Guangzhou. By showing the Guangzhou factories, the fashion show in Paris, and the small tailor shop in Shanxi all in the same film, I hope that we can build up a revealing composite picture.

Rayns: Your film is nothing like a cinéma vérité documentary. You structure both individual shots and the film as a whole very carefully. Does your arrangement of the shots compromise the realities they represent?

Jia: I try to make my shots in a very free way. When I get to the site where we're going to shoot, I try to find the right camera placement very quickly, bearing in mind the need to find the right distance between the camera and the subject—by which I mean the distance that puts both of us at ease and make us feel comfortable. I'm trying to capture the authentic feeling of the space and the people. Sometimes the people being filmed and I form a kind of interactive relationship so that the camera's movements are in sync with theirs. In this light, you could certainly call it an "arranged documentary." But I don't feel anything negative about that description. If we feel confident that we have grasped a certain reality but cannot capture it spontaneously with the camera, why not arrange the elements to reveal that reality? For me, the key thing in making a documentary is subjective judgment. Reality can be distorted by the presence of a close-up camera. It's always necessary for us to feel and to judge.

I'm often told that my essay films are like documentaries and vice versa. When I shoot fiction, I usually want to maintain a certain objectivity in presenting the characters in their settings. But when I shoot documentary, I want to capture the drama that's inherent in reality, and I want to faithfully express my subjective impressions.

Rayns: Did you know much about the fashion industry before you made the film? Was this a voyage of discovery for you?

Jia: I knew nothing at all about the fashion world before I made *Useless*. Making the film has certainly broadened my horizons; you could say that in some ways it opened my eyes. In recent years, "fashion" has become a buzzword in China. The nouveau riche class is wild about such brands as Louis Vuitton, Armani, and Prada. But many people buy these brands because they're famous and expensive, not because they appreciate the designs. And many young people spend way beyond their means in buying these brands, which suggests that ostentatious wealth has become the most important—maybe the only—index of a person's social value.

Focusing on the manufacture and sale of clothes in *Useless* gave me a way of dealing with the changes in China's economy. Like Ma Ke herself, the film cannot avoid confronting China's mania for consumerism. The most interesting thing for me is the way that this mania obliterates all retrospection. Fashion and power are closely complicit in China, and I sense that there's some obscure connection between the mania for newly released brand-name products and the constant erasure of historical memories. That needs to be challenged.

Rayns: Can you be more explicit about the connections between Ma Ke's art and the conditions of the workers in Guangzhou and Shanxi?

Jia: In Chinese culture, the four basic human needs are considered to be food, clothing, shelter, and transportation.

When Ma Ke's Useless label goes beyond the pragmatic function of clothes and directly addresses spiritual questions, I have no choice but to think about the "usefulness" of clothes. In the less developed areas of China, ordinary workers need clothes simply to cover themselves; that gives me the space to connect the clothes themselves with the social conditions of this class of people. I think I'm just using clothes as a medium for looking at society. Eventually, we all have to face up to the conditions in which we live.

Rayns: Is there any special reason why you chose to introduce Ma Ke with a scene in which she's with her dogs?

Jia: What impressed me most about Ma Ke's studio in Zhuhai was the large green area around it and the fact that her dogs could run free. I think that says a lot about her cultural position in the context of China's headlong urbanization. I was also very much struck by the thinking behind the Useless collection: the way it cherishes memory, the psychological implications of the way it incorporates the changes wrought by time. The Useless idea is itself a challenge to China's rapid development and a kind of rebellion. It challenges the obliteration of memory, the over-exploitation of natural resources, and the speed at which all this is happening.

Rayns: Your film is as much about bodies as about clothes. Has your thinking about the body been affected in any way by your collaboration with the painter Liu Xiaodong?

Jia: Indeed, after working with Liu Xiaodong I've started to think about the people in my films more as natural beings than I did before, not just as asocial creatures in a web of relationships. Clothes are an outer expression of our inner world but, as the closest thing to our skin, they have also become symbols of class division within society. When we're naked, though, there is no class difference. All we have then is equality: equality in beauty and equality in our incarnated existence as human beings.

Beijing/London, August 2007.

PART IV
THOUGHTS OF A "FOLK DIRECTOR"

THE AGE OF AMATEUR CINEMA IS ABOUT TO RETURN *(2001)*

30

Translated by Claire Huot

In a restaurant far from Busan's city center, Tony Rayns and I discussed some questions about cinema for the British magazine *Sight & Sound*. The interview was tiring but extremely satisfying. Far away from the hustle and bustle of the festival, we delved into cinema's past, present, and future.

As the sound of the ocean's tide gradually rose outside the window, our conversation approached its end. I don't know why, but discussions about cinema can easily make me a little sentimental. To shake off this mood, Tony changed the topic and asked me: "In the future, what do you think will be the driving force behind cinema's development?"

Off the top of my head I answered, "The age of amateur cinema will return."

This must truly be what I think, because every time someone asks my opinion on cinema's prospects, I find myself repeating it again and again. Of course this answer irritates cinema's so-called professionals. These film professionals, who cling to professional principles as though they were heavenly decrees, who expound at length on their so-called professional team and marketability, lost any ability to open their minds ages ago. They are especially intent on assuring that their own films exude professional refinement. In other words, the images should be as exquisite as an oil painting, the mise-en-scène should be comparable to an Antonioni. They'll tell you how proud they are of the way a flickering spot was aimed just so on the male actor's profile. They are continually mulling over what insiders think, exhorting themselves not to cater to the layman's taste, not to violate the established classical forms. Conscience and sincerity, both crucial to filmmaking, are all but abandoned.

What remains? Rigid concepts and staunchly held preconceived ideas. Such filmmakers are insensitive to new things; they even lack the ability to recognize them. And yet, they ceaselessly advise others: "Don't repeat yourself, you've got to evolve."

Actually, some filmmakers have been for quite a while aware of and on guard against smugness. I think it was at least ten years ago that Krzysztof Kieslowski began to describe himself as an amateur filmmaker from Eastern Europe; and this wasn't casual modesty. From behind these humble words emanated a spirit of independence and self-confidence. As for the recently deceased Kurosawa, all his life he insisted, "I've made so many films but I still don't know what cinema is. I'm still looking for what can make it beautiful."

During this edition of the Busan International Film Festival, Japanese director Kōhei Oguri, a member of the jury, expressed alarm that although the level of filmmaking in Asia had risen considerably in the past ten years—such that it could now be considered world-class—the artistic spirit of cinema had declined considerably. Earlier, during the Hong Kong International Film Festival, another jury member, Wong Ain-ling, said: "Behind the myth of high-cost productions, there is a forfeiture of faith in the culture." It was in this context that the Busan International Film Festival emphasized the importance of independent film in Asia. The ten films entered in the competition were mostly very original works by new names. The festival itself, largely because of this new approach to film selection, has earned the world's attention. In only three years, the Busan International Film Festival has gained much ground on the Tokyo International Film Festival, and for clear reasons.

The focus of this edition of the Busan International Film Festival was "Asian Cinema during the Financial Crisis." Apart from economic issues, discussions focused on how Hollywood had invaded the world, how the same trends traverse the globe, and the danger this poses to national cinemas. During the press conference for *Xiao Wu*, I mentioned how disappointed I was to turn on the television here in Korea only to discover the same cable television content as in Beijing. In a few years, all of Asia's youth will be singing the same songs, shopping for the same clothes, wearing the same makeup, and carrying the same handbags. What kind of world will that be? It's precisely in such a cultural climate that independent cinema, committed to local culture, can provide some diversity. I'm more convinced than ever that it's only by acknowledging diversity that human beings will be able to communicate and achieve equality. The global trend toward sameness threatens to leave us with a drab world. In my talk, I also repeated my conviction that it's always when cinema is in difficulty—when the film industry finds itself in a slump and cultural confidence is at its lowest—that independent cinema, with its critical and self-reflexive independent spirit, becomes a multifaceted creative force to inject new life into culture.

This is why I can claim that the age of amateur cinema is about to return.

This wave of independent filmmakers is composed of people who live for cinema with a passion that can't be reined in. Because they set their sights farther and deeper, their films are beyond the scope of the industry. Their work may be occasionally bewildering, but the emotions they invest in their films find real and tangible expression. Because they ignore so-called professional techniques, they discover more possibilities to innovate. And because they refuse to adhere to fixed professional standards, they are free to play with a diversity of ideas and values. Working outside of confining conventions and practices, they know no borders; they are free. And because they have the conscience and personal integrity of true intellectuals, they are committed and feel a responsibility to others.

Among the archetypes of this kind of artist I count Godard of *Breathless* and Buñuel of *L'Age d'Or*. I would also include Rohmer and Fassbinder, whose admission to film school was turned down.

I know that Polanski once said: "In my opinion, all of the New Wave films are amateurish films." But this arrogant professional filmmaker was ignoring the fact that those talented amateur films brought endless new possibilities to cinema. And that was twenty years ago.

What about today?

Who can say if, among the crowds hanging out in pirated DVD stores, there isn't a Chinese Quentin Tarantino? Who would deny that a contemporary version of Shinsuke Ogawa could emerge in China from among the young people who have the means to play around with digital video? Cinema should never again be the privilege of a few; after all, it originally belonged to the people. In Shanghai, a while ago, I met a group of filmmakers in a club—friends who repaired airplanes or made commercials by day. They may well be harbingers of the future of Chinese cinema. I've always disliked that vague air of superiority that professionals exude. The amateur spirit, on the contrary, is grounded in equality and justice, as well as a concern and empathy for others.

Originally published in *Southern Weekly* (*Nanfang zhoumo*), 1999.

NOW THAT WE HAVE VCDS AND DIGITAL CAMERAS (2001) — 31

Translated by Claire Huot

One afternoon, in a store in the vicinity of the Modern Plaza, I bought two VCDs. One was Eisenstein's *Battleship Potemkin*, the other, Orson Welles's *Citizen Kane*.

At first I didn't think about it, but on the bus returning home, as we passed the field around the apartment clusters by the Big Bell Temple, I realized that for a mere twenty yuan [USD $3], I had two great artists' masterpieces in my pocket. I felt a sudden rush of warmth in my heart, thinking that times had indeed changed. Films that had long been shrouded in mystery—locked away in forbidden archives, only accessible "for internal reference," and strictly forbidden to be copied—were now easily available to the common people. Today, anyone interested in cinema could study these foreign films in the comfort of their own homes. Now we could research poetic montage or depth of field while eating noodles with soybean paste.

In the past, such films were termed "films for internal reference," a term apparently coined during the Cultural Revolution. In the wake of the Cultural Revolution, with calls for the emancipation of thinking, "internal films" were shown more widely. Today, when some films have suddenly been classified "internal," the older generation, who used to have privileged access to "films for internal reference," remember those days fondly and are rather pleased. But how can anyone not seriously lament that the right to watch movies is tied to one's cadre and professional level? This has to be one of China's great inventions. Watching movies was a privilege, which implied one had superior intelligence and moral fiber. "Internal Reference Material"—four words in Chinese: *nei bu can kao*—meant this material was out-of-bounds for ordinary people; only a few deemed worthy were entitled to consult them. Others probably wouldn't understand these movies, or they might steer them onto the wrong path. No admittance. Eventually, a kind of smugness developed among those who were allowed to watch internal films.

Now that things have changed, we can all equally enjoy cinema and share the thoughts and emotions that moving images convey. In Europe, many people don't understand why VCDs, with their poor sound and image quality, are so popular in China. They don't realize that VCDs are inexpensive and there is a craving for knowledge in China because we want to get our rights back. Films that were just titles in books and magazines have become real for us; they are part of our education. We want to see Marlon Brando and Marilyn Monroe. We want to see *Battleship Potemkin* and *The Godfather*. Everyone has the right to partake in humanity's cultural heritage. Just as the dissemination of printed materials in the West once allowed commoners to partake in the knowledge contained in classical writings, the emergence of VCD has broken the monopoly of political connections and work units in China. More and more people will benefit from the circulation of VCDs, ordinary people will be able to view classic films. Just as people who have read many novels sometimes become novelists, we can imagine some who might become filmmakers after viewing VCDs. Moreover, in the filmmaking business, many individuals have become

NOW THAT WE HAVE VCDS AND DIGITAL CAMERAS (2001)

outstanding talents in part because they have devoured a great many movies. Once, you had Godard and Truffaut, a couple of social slackers who spent their days buried in the film archives. More recently, you have the video rental shop clerk, Quentin Tarantino. These are examples of people who succeeded, and their success stems primarily from their insatiable and constant viewing.

Which brings me to the question of digital cameras.

The smallest digital camera is the size of your palm, but you should absolutely not underestimate it. This digital camera and video recording device may cost only RMB ¥10,000 [USD $1600] or so, but it can shoot extremely high-resolution images. Abroad, more and more people use these to make films, especially documentaries. Here, more and more people are seen with that gadget in hand.

One day, in a bar on the east side of town, I came across four film crews shooting a documentary. Although I was puzzled that four directors were filming the same rock band, I sighed with satisfaction to see that they were all using high-quality digital cameras. Indeed, the appearance of the digital camera has made film simpler, more versatile, and inexpensive. Affordable and easy to manipulate, this camera allows a greater number of people to express themselves with moving images. As the Super 8 home movies trend did years ago, digital cameras put filming in the hands of the individual rather than the hands of the industry. This practice has the potential to transform Chinese cinema. For example, more and more people use digital cameras to make documentaries and experimental films. Documentary filmmaking presupposes an altruistic spirit, while experimental cinema requires a spirit of innovation—two things presently lacking in Chinese cinema.

In Europe, especially in Switzerland, some companies specialize in the transfer of digital images to celluloid. Although some twenty thousand RMB [USD $3200] may seem expensive, such a transfer allows digital films to be shown in movie houses and assures a future for digitally filmed works. While I was just learning about this, outstanding practitioners from Europe had already made their mark. A few filmmakers from Denmark published the Dogme 95 Manifesto. They opposed the film industry's stifling of creativity and advocated artisanal filmmaking with a minimal use of artificial lighting, a modest budget, and the use of only a handheld camera. Their successful works include the hit *Breaking the Waves*. Digital technology was ideal for Dogme 95's purposes.

I saw Wim Wenders's documentary *Buena Vista Social Club* in a movie house in France when it had just come out. This film, mainly shot in Cuba, depicts the lives of a few old jazz musicians. It was also shot with digital technology and then transferred to film. The coarse grain of the images on screen produced a documentary-style aesthetic, and the shooting possibilities of the digital camera allowed the filmmaker to capture powerful images. The audience applauded wildly throughout the entire screening. I couldn't help but sigh in admiration. A new film aesthetic, conceived with digital technology, was taking shape. The digital camera, with its low requirements for lighting, its small size, easy operation, and low cost, offers great prospects for the future.

VCD gave us the opportunity to see all manner of good films; now that we have the digital camera, we can easily film moving images.

Put a digital camera in the hands of someone who has viewed large quantities of VCDs and there's no telling what he or she can do. The invention of film, then the invention of VCDs, and now the invention of the digital camera—those are three inventions for which we can be truly grateful.

Originally published in the Film Section of 163.com, *NetEase* (*Wangyi*), January 21, 2001, http://movie.163.com/edit/001214/001214_67256.html.

HAVE YOU BOUGHT JIA KEZHANG'S *PLATFORM*?

32

Translated by Claire Huot

A few days ago, I wandered into a shop selling pirated DVDs in Beijing's Xiaoxitian District. It was noon and the store was quiet and deserted, just the shop owner and me. I clambered up on a cardboard box and eagerly searched for a long time without finding anything.

Seeing how persistent I was, the owner spoke to me, "It's because of the 16th Congress,[1] there's little to no new stock."

I immediately realized I knew much less about politics than the owner and that I had better leave.

The owner must have had a sudden inspiration, because, right as I was reaching out to push open the door, he blurted out to me: "I've got *Platform* by a Jia Kezhang.[2] You want it?"

I was stupefied and replied: "What?" The owner repeated his words, I faked indifference and asked: "Where?"

The owner said: "I'll have it tomorrow."

After I left the shop, my heart was beating wildly. Like a father who has lost his child and unexpectedly discovers him in a human trafficker's haunt, I felt excited but also depressed. During the evening I couldn't calm down. I was either calculating how many people would be seeing my film, and couldn't help

---

1. The 16th National Congress of the Communist Party of China was held in Beijing between November 8 and 14, 2002.
2. The jumble of the syllables in Jia's name is also funny when written: Jia Ke Zhang means "Sham Department Head."

but feel proud, or I was remembering the hardships of shooting this film that people had now stolen, and I'd become unhappy. The problem with this generation is that we worry about personal gain and loss. That goes for me too. I can only overcome it gradually. But that night went by fast. At approximately eight o'clock in the morning I woke up naturally; I usually sleep in but, oddly enough, I was wide awake. I took a cab to Xiaoxitian and bought *Platform*.

Back in the office I watched *Platform* again, three years after making it. This gave me some distance. Robert Bresson has said that each film has its own life: once it's shown to the public, it no longer has anything to do with the filmmaker. You can only wish it good luck. Still, watching *Platform* brought back to mind a lot of events of the past. And those events are the reason why I chose to make films.

I saw the ocean for the first time when I was twenty-six years old. The first thing I did after learning how to ride a bicycle was to ride my bike to the county city over nine miles away to see the train. These things are now happening to my protagonists who are ten years older than me. At that time, for a child like me who had not stepped outside his small county town, the railroad meant faraway places, the future, and hope. I've experienced the wild yearnings for the outside world that permeate *Platform*. I remember that when I was a seventeen- or eighteen-year-old schoolboy, I could never fall asleep at night. I was forever looking forward to the next day, because I felt that the new dawn would bring new changes and, consequently, new things would happen. This feeling is still with me; people who have had experiences similar to mine probably feel that way.

I first studied fine arts. In those days, our study of art was not at all romantic. The pursuit of art was not our goal; rather, art was a way to get out. In the county town, if you wanted to go live in bigger cities, there were two outlets: you became a soldier or went to college. For me, becoming a soldier was not an option. I could only apply to university. But my grades were very poor, so I opted for painting because culture courses in fine arts schools were not demanding. That's the reason why a bunch of us went to art school. At the very beginning, we had no goal other than eking out a living. Actually only one or two of us passed the higher education entrance exam. The others who didn't pass in the first year went back home, and those who tried again in the following year but who still didn't pass gave up. Finally, I passed the entrance exam to the Beijing Film Academy. At first I thought I was really something: "Look at me! I persevered! I pursued my ideal!" But once I was a bit older, I realized that it's much harder to abandon your ideal than to pursue it.

At that time people abandoned their studies for various reasons: maybe their father had suddenly died or their family needed a male to work. There were also families who couldn't afford it and didn't want to use up the family's savings. Each one had a very specific reason; each one had to assume some of life's responsibilities. They abandoned their ideal because of their sense of obligation to others. In contrast, those of us who allegedly persevered in our pursuit actually paid a much smaller price than the others. They had to take up a very ordinary, routine life. They knew the consequences of abandoning a dream, but they still abandoned it. Small-town living today and tomorrow will always be the same; a year ago, and a year from now, there's no difference. This film is sad because life for these people stuck in that place promises no miracles or possibilities; all that remains is a mediocre life in a fight against time. Once I understood this, my attitude toward people and events changed radically. I started to truly commiserate, to get really close to the so-called losers, the so-called ordinary folk. I felt I could see that they had strength and that their strength is the force society leans on to move forward. I put all these emotions and thoughts into my film; I tried so hard to talk about our lives. But did people come to listen?

The owner of the music and film shop is still hawking away; it's as if he's helping me to ask: "Want Jia Kezhang's *Platform*?" I don't want to correct his mistake, because I'm now in a pretty good frame of mind.

LETTER TO THE YAMAGATA INTERNATIONAL DOCUMENTARY FILM FESTIVAL

33

Translated by Alice Shih

When I was young, I liked to stand and watch the people on the streets. The unfamiliar faces that passed in front of me gave me a strange sense of warmth. I wondered if they were living lives similar to mine: their rooms, food, stuff on their tables, their loved ones… Might we have similar problems? I'm afraid that I might lose my curiosity to know other people's lives, so I make documentaries to help me conquer this fear.

It seems easy to lose our sense of connection with one another. There are usually just a handful of people in our lives. But documentaries can expand our lives and eliminate our loneliness. For me, the most important thing in making documentaries is to capture the spirit of justice and courage, which I fear could leave me. Through documentary filmmaking I get to feel that every life on earth is dignified, including my own.

Film is a form of recording, and documentaries can help us retain traces of our lives. It's one of the ways we can fight against oblivion.

This article was written for the program book when Jia served as the international juror of the Yamagata International Documentary Film Festival in 2005.

34 FIREWORKS THUNDERING BUT VCR BLUNDERING

Translated by Alice Shih

In 1999, when I was about to shoot my second feature, *Platform* (*Zhantai*), I put some locations and casting information on DV tapes, transferred them to home video format, and sent them over to one of my producers, Kitano Takeshi. Coincidentally, I was invited to be a juror for the upcoming Busan International Film Festival, so we had decided to meet there. Almost every film festival in the world has a video library for industry guests to view tapes should they miss a screening. Producer Ichikawa Shozo, Mori Masayuki of Office Kitano, and I went to the video library to play my tape on a VCR. The lady at the service desk was very polite, but the machines were not cooperating. The first VCR was on strike; the second player worked, but it couldn't play PAL system tapes. So there were phone exchanges, and Korean men of different shapes and sizes kept entering and exiting the room. One of them was sweaty, one kept yelling at the machines, and another one, who really took his time and appeared to be rational, came and left quietly. At the end, the service desk lady smiled and apologized to us nicely. Disappointed, we helplessly looked at all the different video players of various ages and left.

That was the year when the Busan International Film Festival generated a lot of buzz, overtook the Tokyo International Film Festival, and was the rising star in the festival world. The festival had lots of funding and was able to invite distinguished guests, putting them all in limousines and five-star hotels. On the other hand, the hardware of the festival operation seem to be lagging behind, hampering Busan's vision to become the Cannes of Asia. At the closing ceremony, winners were honored with spectacular fireworks over the stage. Watching this dramatic moment above me, I thought and estimated the number of VCRs they could buy with the money spent on the fireworks. But these dazzling fireworks could become news headlines, and it is very unlikely that a television crew would do a broadcast from the dim video library. There is no contest between fireworks and VCRs. This made me want to distance myself from this festival.

Recently my peers in Korea were demonstrating on the street about film import quotas. Having a passion for films and a duty to defend one's own culture is certainly respectable. Yet, too much protectionism can have a negative effect on cultural confidence, and quota systems aimed against Hollywood dominance is like a cold war that will only make both sides uncomfortable. I heard, "Film is a unique product, it needs protection!" This argument sounds valid, but sacrificing the free trade principle to market protection is not the most legitimate way to go. The French government and filmmakers have better ways of protecting the French film industry. They levy hefty taxes on Hollywood distributors to subsidize local film productions. They also reinvest a portion of their television commercials' income into French films. These methods prove to be far superior. They increase the competitiveness of the local film industry without excluding foreign films.

Last year, English film critic Tony Rayns wrote an in-depth article on Korean films in *Film Comment* magazine. About *3-Iron* (*Bin-jip*) (2004) by Kim Ki-duk, he wrote, "For someone who hasn't seen Tsai Ming-liang's *Vive L'Amour* (*Ai qing wan sui*)

(1994), *3-Iron* might get some attention; otherwise the film is just another strange cliché."

I agree with him. Please don't always look to Korea for Asian contemporary film classics.

Originally published in *Southern Metropolitan Weekly* (*Nanduzhoukan*), March 13, 2006.

THIS YEAR WILL EVENTUALLY COME TO AN END   35

Translated by Alice Shih

Winter arrived early this year. The scenery turned bleak by October. I used to like winter. Wilting flowers, drying willows, and the falling of leaves gave me insight into the universe. The color gray seemed strong to me. But this year the cold really got to me. The afternoon sun brings less warmth into the room. The occasional metal clinks heard from the distant boiler room don't resonate as before.

This has not been a productive year for me. I let my work stop, and I haven't shot one frame of film. In fact, I feel like I have lost my audience. Life is spinning and my films are powerless. The decadent habits of my youth have come back to haunt me. Now I put on shows instead of doing real promotional work for the media. There is a lyric that I like:

> *Who would care about our lives?*
> *If hardships were all we saw*
> *And our remarks were taken as complaints,*
> *I should stop and distance myself from my work.*
> *I should stay away from Beijing.*

Mount Wutai[1] was wrapped in snow when I arrived. Nature doesn't care about the mood of her viewers. Mother Earth has her

---

1. Mount Wutai is a Buddhist sacred site, home to many important temples in the northeastern province of Shanxi. It is made up of five flat-topped peaks, including the highest point in northern China. The area was inscribed as a UNESCO World Heritage Site.

own logic for the weather—snow, rain, or shine. This is the privilege of her nobility. The leisurely pace of the seasons is not altered to please any audience. You have to accept what she does; you cannot change things to suit yourself. It is like watching a Robert Bresson film. You must enjoy it the way it is; it will not bend to please you.

I'm glad this pilgrimage has given me some inspiration for film ideas. The wilting grass in Beijing is getting a coat of fake green spray, but the ocean-blue color on the apex of Dai Luoding Temple looks very real. Crying out from the mountain slopes, I hear echoes down the valley. Nature is teaching me.

I got back to Taiyuan and called some of my friends. I hadn't heard their voices for a few years. Another lyric comes to mind:

> *We were young once and loved to dream.*
> *We only wanted to fly ahead.*

We used to hang out every day, but I have grown apart from these brotherly friends—too busy pursuing fame. After a few drinks I saw that time didn't change us that much. They called me by my nickname, and we started talking about things that we wouldn't discuss with others. They asked me to consider having a kid. They worried about me in my old age. I felt like crying. It seems like you can expose your weakness only in front of your true friends. They don't care about films; they just don't relate to them. They only care about me as a person; how I live my life is important to them. I don't get to experience this kind of warmth very often. As a director, I have to stand strong, not afraid of tomorrow. This drunken wildness with old friends was like a spark in my humdrum life. I threw up and said, "I love Jianghu."[2]

---

2. *Jianghu* is a term derived from swordplay (*wuxia*) stories. It often refers to anarchy, the lawless world of the gunslinging Wild West. It is often used to describe the modern criminal underworld but can also describe a community of common interest where they have their own code of conduct.

I carried on westward to Yulin. The old guy sitting next to me on the bus was very quiet. He suddenly asked me the date as night fell. I told him we were approaching the end of the year and he sighed, "This year will eventually come to an end." I have no idea what kind of problem he was facing in his life, but if he were so eager to see the end of it, I could tell that it must have been very rough. As with my films, which don't have much plot, you can get the hints that surface from these traces of life.

Life has taught me another lesson, and I have to believe in film again.

DARKNESS AND LIGHT OF 2006

36

Translated by Alice Shih

1. Before the theater

It was late autumn in 2006; I went to Datong by myself. I arrived at the station around midnight and it looked like snow was imminent. The traffic flow on the platform was busy. I took my time to disembark so I could savor the joy of being alone on the platform after the crowd dispersed.

I didn't ask my friends to pick me up. I hopped in a cab outside the train station and headed to a small restaurant. The city seemed to be asleep already; only the west side of the streets was lit up, and all was dark on the east side, waiting patiently for the morning sun. I got out of the cab and saw that lights were still on behind the green door of the restaurant. I pushed the door open and found my friends at a drinking party. I knew I would find them here; they had nowhere to go in this town other than their nightly party routine. I didn't even need an appointment.

We didn't want to go to bed after drinking, but we couldn't think of where we could go. We finally hired three cabs and the twelve of us went to a teahouse to play cards. A waiter dressed in a traditional Chinese top brought us a few packs of cards. I made a small bet and won a little, using what I learned about cards when I was in high school. Suddenly we had a power outage and all went quiet. Twelve cigarettes lit up the loneliness of the city. I felt a sense of melancholy watching their red ends glow.

When we got out to the street, the only lights we saw were from a gas station and a sauna. The hands of strangers would send us to sleep tonight. The heat was on high inside the massage room. I opened my eyes wide. In warm and quiet places like this I can let myself go to a place of emotional tenderness. But I had become a man of the world, as they say, and so I did not weep.

Like me, all the other men of the world woke up around noon, all wearing cheap shorts and tank tops. We gathered in the big dining room eating mutton noodle soup. Our cell phones rang one after another in the locker room compartments as we changed, and I got to hear how everyone dealt with their wives differently. After we all gathered outside the sauna house, as if it were the New Year, we looked at each other and asked, "What's next?"

2. Inside the theater

We had nowhere to go and decided to split up into a few cabs to hang out in the mining area. More than a dozen coalmines littered the Datong highway, each assigned a number. A building with Eastern European architecture could be seen among the greenish gray structures of the mining district. It stood out with pride but seemed underappreciated. We got closer and found out that the building used to be a movie theater. The fading star at the entrance was a remnant of that era's idealism, yet now all the doors and windows were blocked up with bricks, displaying nothing but loneliness.

Entering the dim and empty lobby, I didn't see any seats. Instead, I saw how the ceiling decoration was an imitation of the Great Hall of the People in Beijing—an emblem of its past glory. Today this theater serves as a warehouse for cement.

Miraculously, I found a strip of 35 mm film. Unspooling it under the light, I saw the title of the film: *Blood Is Always Hot* (*Xue zong shi re de*) (1983).

This title reminded me of the reformist image created by the lead actor Yang Zaibao and the heydays this place had seen. The silver screen had since been taken down. In this big piece of land, the fate of theaters is like that of old temples. Hibernating, they either wait to be demolished, or, perhaps far in the future, they may be exalted as historical sites. We roamed farther down the road hoping to find more traces of theaters or clubs. We did find more theaters standing against the wind and dust. Some had all their glass shattered, like a body covered with wounds. Some stood silently, witnessing the constant changes of human time. I suddenly realized that the faces of contemporary Chinese people, their joys and sorrows, reunions and separations, were not exhibited on these screens. Not far from here were the miners' quarters. What caters to their emotional needs?

Film should be an affordable entertainment for the masses. In old Sicily, average people could sit and smoke and witness Federico Fellini's surrealism. Now community movie screens are taken away from the common people. What they call the "general audience" is in a far-off metropolitan city, and they pay fifty RMB ¥10 [USD $8] to watch a movie. The people who live in the wilderness, like the miners here, cannot afford it. Without these local community theaters, they can never become a part of this misnamed "general audience."

3. Film
It's getting dark. My brotherly friends and I are going to party again. We'll do the same things as always, carrying the same solitude and emptiness.

This is our state of mind. It's the reason why we make our films the way we do. Going through the contents of *Chinese Cinema 2006* with open eyes is a way to understand the mental state of our nation. We need to make more films; we need to experience more moments of darkness and light.

Originally published as the preface to *Chinese Cinema 2006* (*Huayu dianying 2006*)

BEWILDERMENT   37

Translated by Alice Shih

The date was January 13, 1999, when I was summoned by the Film Administration Department in Beijing. I was twenty-nine years old. I had just graduated and had never entered any government agency before. I anxiously searched east and west and finally, in the East Four Alley, saw the black words written on a white signboard: "State Administration of Press, Publication, Radio, Film, and Television." I was about to go in when seven or eight middle-aged men emerged. I recognized one of them, so I quickly stepped aside and watched. He was a well-known Fifth-Generation director, and he was embracing a government official like a brother, while others laughed and joked outside this elegant and stately heritage building. I was a bit bewildered; I couldn't imagine how this film master could be so at ease with bureaucrats and not feel intimidated.

The crowd of men dispersed as exhaust shot out of the tailpipe of the director's jeep. In the quiet alley, I beat myself up for being a rookie. The officer wasn't that scary looking; in fact, his scholarly face resembled an aged Winston Chao, the actor.

I walked into the big courtyard and heard the guard calling for me to stop. I got startled and a bit nervous. I explained myself and he gave me directions. I went through corridors and pillars and was about to knock on a door when it opened and out walked the "Older Winston Chao." Coincidently, I found out that he was the one I spoke to on the phone to make this appointment. He explained that he wasn't the one who would talk to me but that he would show me the way. He told me that this building was the former residence of Prime Minister Liu Luo Guo.[1] I suddenly thought of the funny face of the actor Li Baotian, who played him in a TV series, and I broke into a laugh.

We sat down and he served tea. He asked me to make myself comfortable and to wait while he stepped out of his office for a little while. I looked around his office while he was gone like a full-circle camera pan. I saw a document on his desk with my name on it. I knew I shouldn't, but I couldn't resist. Since there was no one around, I picked it up. It was a copy of the *Dacheng Newspaper* (*Dachengbao*) from Taiwan. The film and entertainment section had a featured article on my film *Xiao Wu* (1997). What surprised me wasn't the article itself but a handwritten note on the page, alerting the officials not to let films like mine influence cultural exchanges outside of China.

I was reasonably upset. And it grew to rage and hate. After I got hold of myself, I discovered this little report was signed by XX. XX happened to have been a member of the crew of that Fifth-Generation director that I saw earlier. I couldn't believe it! Did I step on your toes somewhere? We're in the same profession, why are you trying to make my life difficult? We should be courteous to others! Why are you bad-mouthing your peers? I was bewildered! I put the article back in its place and sat still in my seat. I heard myself sighing as I burst into tears. I wasn't crying for myself but for who had written that report. I thought of what Romain Rolland had written: "I have nothing but much sympathy and pity for them today!" At that moment, I felt that at least I had the upper hand in morality.

---

1. Liu Luo Guo (1719–1804) was a politician, scholar, painter, and calligrapher. He served as prime minister in the court of Emperor Qianlong in the Qing Dynasty.

"Older Chao" seemed to be in a good mood as he walked in. He asked me if I knew why I was summoned. I told him I knew. He took the article in his hand and handed down the sentence: "From this day on, Jia's right to shoot films has been suspended." We went quiet. He picked up the article from his desk, dropped it down heavily, and said, "We didn't want to discipline you, but your peer and elder told on you."

It was as if I was sleepwalking when I left his office with the disciplinary document in my hand. Pacing the alleys alone, through light and shadows, I realized that people's hearts are so mysterious, complicated, and unfathomable. Bemused, I thought, "If I hold on to the memory of this bewildered experience, it might be able to calm me down in the future."

Originally published in the *Soho Post*, May 9, 2007.

THE COWARDICE OF OUR ENTIRE GENERATION   38

A Lecture at Peking University
Translated by Alice Shih

Hello! Thanks for finding the time to come to see *Still Life*. I was sitting in the audience with you just now. I planned to watch only the first ten minutes, but I ended up staying till the end. I was still shooting this film in May, and it seems like I have already forgotten, after only three months, what happened. Watching this film, I found it unfamiliar and, at the same time, very recognizable. I spent almost a year working with my coworkers, but we humans are forgetful; we have poor memories, and that's why we need films.

  I went to Sanxia (the Three Gorges) thanks to Liu Xiaodong. I was there shooting a documentary on the artistic world of his paintings. I have loved his paintings ever since I saw his show in 1990. He is able to uncover the unnoticed poetry in everyday life. This poetic flavor exists around our daily activities. The idea of a documentary on him was on the back burner until last September, when he said he was going to Sanxia to paint eleven workers. I decided to follow him and start shooting my documentary, *Dong*.

  We could have chosen to be typical tourists at Sanxia, admiring the scenery of the ageless mountains alongside the flowing water, but once we got on shore and started strolling on the streets, entering the neighborhoods and homes, we discovered that there are actual people living in this ancient landscape. These people are very poor and under tremendous pressure. This huge government project is moving a million residents, demolishing a thousand years of civilization. It has brought rapid change and placed incredible stress on their shoulders in the process. Tourists like us bring our cameras, relaxing as we tour the scenery, as if their difficulties have nothing to do with our lives. Yet, when we pause to reflect, this tremendous pressure could also be ours. What if our daily morning subway rush produced such helplessness, such loneliness? Almost every Chinese resident in every province has been subjected to this rapid modernization, which supposedly brings us material satisfaction in return. Entering a supermarket today, we bask in material abundance. At the same time, we are shouldering the pressure exerted on us by this modernization. We too experience the spatial alteration, the sleepless nights and the day-for-night lifestyle, just like the residents of Sanxia.

The moment I stepped into the region, I felt the humidity. I stood there looking at the pier. Fengjie is a busy river port, harboring people of all walks of life who wheel and deal and exhibiting the ongoing hardships that Chinese citizens have to endure. I got my desire to shoot right at that moment. I began shooting my documentary on Liu Xiaodong and gradually entered the world of his subjects. That day, I was filming Liu and an old man, one of his models. I found him very comfortable in front of the camera. (He later became one of my actors; I had him hand ten yuan [RMB] to my main character, Sanming, in *Still Life*). As he was walking out of frame, I spotted him taking a puff of a cigarette with a sneer, as if he was upholding his dignity as well as expressing his disapproval of the film—commenting on us tourists with our ignorance of the locals' lives. I couldn't sleep

THE COWARDICE OF OUR ENTIRE GENERATION

a wink that night at the inn thinking about that possible fatal limitation of documentary filmmaking: every person's natural self-defensiveness.

I started getting really creative with the storyline. I visualized what hardships the locals must face, as well as the concomitant stress. Very quickly, the story of *Still Life* took shape. I spoke with my assistant director about telling this story from an outsider's point of view, since we could never understand the great transformation in the same way as the locals who went through the changes from day one. We would depict the region from an unfamiliar angle. The river has been flowing for thousands of years, and countless individuals have passed through. This high-traffic region has been progressively corrupted and has grown into a Jianghu-like lawless community over time.

Who isn't living in a harsh community these days? You might be a reporter living in the ruthless corporate world of media or a real estate agent struggling in a tough market. You have to follow certain rules in your circle and fight to stay alive in your treacherous community. In the film, a buck fifty will get you a night in an inn where the landlord fights the same battle to make beds and thus his livelihood. Like this innkeeper, he needed to take such actions to live. With these thoughts bombarding my head, I rushed to write the script.

I ran into a singing kid on the street. He took my hand and asked if we needed a place to stay. I told him no, and he immediately asked if I needed to eat. I told him we ate already and he looked so disappointed. He then asked if I needed a ride. I turned to ask what business his family is in. He just smiled. I watched this fourteen-year-old boy walking away, thinking that this enthusiasm might just be the appropriate attitude to have in life. I found him again later and I asked him what he liked to do the most. He said he liked singing, and he sang Mice Love Big Rice (*Lao shu ai da mi*) and *Two Butterflies* (*Liang zhi hu die*) for me. I asked if he knew any songs by Teresa Teng. He couldn't do any, even after a lot of coaching; he only knew *Mice Love Big Rice*. So we used this song in the film, and his appearance was like an angel.

Under all circumstances, people try to keep their dignity, and they try to survive. This idea is essential to my characters, including my two leads. I immediately thought of my cousin, the son of my second aunt. He acted as a miner who signed a waiver on his life in my film *Platform*. He also played the relative who carried a black bag to handle the funeral arrangements of the character Er Guniung in *The World*. I thought he should be the lead this time. I was very close to him when we were young. He left to work in the mines when he was about eighteen years old and we grew apart gradually. I know that he has heaps of passion buried in his heart, despite his reticence. We didn't have much to say to each other any more when he came home; he was so distant and unfamiliar. We just smiled at each other. His face kept popping into my head as I visualized this film. And I realized why I kept making the same kind of film and why my camera kept dwelling on silent faces like these for the past ten years.

We are too comfortable living in our own boundaries, thinking that our world is shared with everybody else. We need to wander outside, to look closer at our relatives. We'll realize that it is not the case. I believe we need to go on shooting so that we don't forget the real world so easily.

My cousin came to work as our actor, and he did a terrific job. At the beginning, I was particularly worried about his interaction with the local actors, since there was a language barrier. He reassured me and said, "Brother, you don't have to worry, I understand them. I've worked with many miners who came

THE COWARDICE OF OUR ENTIRE GENERATION

from Sichuan and the neighboring provinces." He communicated with the local actors very well. He did especially well in the scene with his character's ex-wife by the river. She asked him, "It has been sixteen years, why did you come to look for me now?" In my script he replied, "In spring, I had an accident and was buried in the mine. When I was under, I thought to myself, if I ever get out, I will look for you and the kid." That was a good take and we cut. But he pulled me aside and asked if I could do another take. He said he didn't want to say his lines. He didn't want to confess the reason. We all know what happened inside the mines; some things are better left unsaid. It would be more emotional without words.

His suggestion was excellent. There are so many things in life that don't need to be explained clearly. There is no need to spell out every cause and effect. We hear so many stories these days. If we are open, we can jump out of our own narrow world to observe and understand the lives of others.

Perhaps we had a similar difficult experience but chose to forget about it. When we are alone, and have the courage to face and deal with the experience of loneliness, we can empathize. At other times we are not strong enough to face our own memories or even watch films that remind us of them. This is the cowardice of our entire generation.

The locals there don't talk about "job hunting." They say: "strive to survive." Likewise, I think we should have greater courage to accept whatever comes our way. Hunting for a job is indeed an act of survival. They are not apathetic but optimistic. I felt my body reacting to this warm local blood while I was shooting. It burnt inside me and I found the courage to face myself.

Then I started to shape the character of my leading lady, Zhao Tao. We had worked together a few times, and she would be playing a woman married to an absentee husband. We shot the night before she had to make a decision about her estranged husband. The original script described her as being alone, yawning, and clueless. I chose a documentary-style shoot. I shot her sitting around for more than an hour; she really got tired and impatient, until she finally fell asleep. When I was ready to wrap, she asked me to look at the fan on the wall and said, "Breaking up is a big and difficult decision. To express her wavering emotion, can we shoot her using the fan to blast away the Sichuan heat and humidity? To relieve the anxiety inside her?" So we shot her opposite the fan, as if dancing with it. It looked to me like the dance of a common person, an everyday individual. Even a woman who hurries on the street possesses her own beauty. I felt I had shown this type of beauty with the help of an actor's creativity.

After that, we worked hard on the film. People came in and out; rain or shine, we kept on shooting. When we were done and the whole crew moved back to Beijing, none of us felt at home. We couldn't get adjusted to the crowds, the pace, and the hustle of the city. It was like we had left behind our reserved happiness, compassion, and memory in Sanxia.

Now that the film is finished and ready for release, we have chosen to show it between the 7th and the 14th. There isn't much time, only a week to receive its audience. We're prepared to "dance" our best for these seven days, to let *Still Life* reach out to an audience of "Good People" (the direct translation of the Chinese title of *Still Life* is "The Good People of Sanxia"). This is not a rational decision, but I want to see who really cares about "Good People" in these money-grubbing times. [Long applause]

Earlier today, when I was on the road here, I saw the same faces rushing to and from work at dusk. I felt like I was feeling the same old humidity all over again. But I don't feel melancholic this time. Perhaps I still have a dream in me, and it hasn't been destroyed yet. Thank you very much! [Applause]

Lecture at Peking University, December 4, 2006.

BLOCKBUSTERS ARE LIKE A CONTAGION AND DESTROY SOCIAL VALUES   39

A Conversation between Xu Baike[1] and Jia Zhangke
Translated by Alice Shih

WHO WILL DEFEND FILMS?

Xu Baike: You seem to be more vocal these days; you've even been a bit sharp at times. The general impression we got from you previously was more low-key and modest. Do you agree?

Jia Zhangke: Yes, you are certainly hearing more from me. All directors, including me, are facing the many problems created by the entropy of the Chinese film environment. The myth of commercial cinema shrouds the whole industry. We don't have access to truly valuable Chinese films, since they don't have a chance to reach and be recognized by the public. This is the biggest problem. Some people disagree and put me on the spot, saying that my visibility and speaking platform count as distribution and publicity for my films. Is this true? I think they are mistaken. I have to agree that I have been given

---

[1]. Xu Baike, reporter and editor of the *Chinese Youth Daily* (*Zhongguo qingnian bao*), *Ice Point Weekly* (*Bingdian Zhoukan*), and author of *Those Chinese Republicans* (*Minguo Naxieren*).

many commercial opportunities. Whether I do well or badly at the box office, my films can still reach an audience on the big screen. Fortunately, I am able to survive in the current environment, but for many emerging directors the prospect of survival is grim. There don't seem to be theaters to accept their work. Often it is so tough that some of them have to abandon their art and find work outside of the film industry.

Xu: Why is that?

Jia: Some good films are shot on DV tapes, but technical requirements for theatrical release are either 35 mm film stock or the costly high-definition format. Not even 16 mm films can be shown in theaters, so DV doesn't cut it.

To set such a high standard for technology really isn't necessary. We all understand that directors should be given the freedom to choose the formats they prefer, and the theater owners should determine if they will accept these materials according to the market demand, not according to the current system of state dictates. It is not fair. This practice has shut many directors out of Chinese cinema. They could be marginalized forever.

We have been voicing our concern about the situation, but there has not been any response. There are many influential and vocal directors in the industry but no one has stepped forward to say anything. They just watch these young directors who fail to get a chance to screen locally start to find audiences at film festivals.

Then they turn around and accuse these emerging directors of making films just to please the festival circuit and not the general public. This really isn't fair.

When I first started making independent films, the media was quite supportive of this form of creative work. They used their resources to promote us. Yet, in the past couple of years, since all these big blockbusters have prevailed, values have changed and we are hardly able to find any media coverage of emerging directors and their work.

Society is not giving much of a chance to new or burgeoning directors. This has been going on for almost two years. I couldn't hold it in anymore and I started speaking out at the end of last year.

What's paining me is that directors more influential than I fail to take any action. Why is the Chinese film-rating system still not ready? Why isn't anyone bringing up the feasibility and flaws of the review system on a bigger platform? Prominent directors holding official titles could sway government policies, but they never promote reform. I was fantasizing that someone more qualified than me would assume this responsibility, but I was disappointed. I have never taken myself as somebody significant or persuasive, but what I can do is rant.

THE "GERMS" CARRIED IN THE BLOCKBUSTERS

Xu: Personally, I really enjoy your films, but I have to confess that I didn't watch *Still Life* in a theater. I bought myself a DVD. I did spend more money for a ticket to see *Curse of the Golden Flower* (2006) in the theater. As far as I know, many of your supporters bought your DVD instead of going to the theater.

Jia: Are you trying to ask me why things turned out this way?

Xu: Yes. This seems to be happening a lot. We stepped out of the theater and started cursing the blockbusters, but we didn't support the films that we really liked at the box office.

Jia: What we are facing today is what we call high input and high output. The concept of greed associated with movies is entrenched in the people. The influence by word of mouth becomes a new phenomenon, and that applies to the mythical influence of blockbusters.

I have to make it clear that I have nothing against blockbusters, and I am in no way anti-commercial cinema. I'm just calling on the Chinese commercial film executives to make changes. I do not criticize blockbusters because they require huge budgets. I understand that filmmaking is a business. If you can find the money, it doesn't matter how much you put in or how much you make in return. What I'm questioning is their business ethics. They operate like fascists, and they are destroying what many of us hold most dear.

Today blockbusters in China succeed by flouting basic social principles. Let's take the principle of equality. Chinese blockbusters dominate the theatrical screens, and they partner with the authorities to occupy all public resources. In this environment, non-blockbusters become obsolete. Big executives begin by taking over China Central Television new waves, as well as all the other channels, to broadcast that a certain blockbuster is opening soon. The way they mobilize public resources is incredible; they occupy every airport billboard and they dominate television and newspaper advertisements. When all social resources are gathered to promote a film, this film is no longer merely an entertainment choice; it becomes a public affair. I don't think I am exaggerating when I say that this fascist dominance is sociologically dangerous.

For the public it is a duty. They go have a drink, do some knitting, chew melon seeds. They can predict what will happen but go anyway because they can't miss out. This is how blockbusters work. Some will go to criticize them and others go despite the bad reviews so they can join the conversation. This process happens over and over again.

It saddens me to think that film promotions depend on practices that are counter to social equality and democracy. But another threat is the values that these blockbusters promote. They prioritize entertainment only, and they denigrate the function of film as a vehicle for ideas.

I remember when everyone was criticizing the script of *House of Flying Daggers* (*Shi mian mai fu*) (2004). Zhang Yimou defended himself, citing that the film was for entertainment. The public was supposed to go into the theater just to have a laugh. When *Hero* (*Ying xiong*) (2002) was criticized, he questioned why films should be laden with philosophies. I don't think the question was about *Hero* not being philosophical enough. It was that the film preaches a hateful philosophy that we would like to fight against. But when he was criticized, he denied that he was transmitting any kind of philosophy! The film has a lot of discourse on the nature of the world. What is that if not philosophy? How can he say it is mere entertainment? This act of dissimulation shows us that he no longer has the temperament to confront this question as a serious cultural topic, and that is terrible!

Producers of blockbusters stress that the public has a choice and things are determined by market demand. But these choices are controlled by powerful executives. The public really just goes with the flow. I wonder about the number of moviegoers who actually make an independent judgment call. I think culture should help the public cultivate critical thinking so that people can make up their own minds.

Blockbusters are all about entertainment for profit. They denigrate and negate films as vehicles for ideas. This affects the people. The public is told that they are painfully tired and what they need is entertainment. This is unethical. Why should anyone be allowed to watch films by Jia, he just portrays the dark side of life?

Xu: I could tell you what a coworker once told me: "Jia's films are too close to reality; they mimic real life. Life is tough already; why would we want to go into a theater to watch more harsh realities? Aren't films supposed to be dreams?"

Jia: Are we, in fact, trying to disable the power of arts and culture? Why is tragedy the main genre of culture? Why does the human race need tragedies? Do you really want me to answer these basic questions? I really like what Liu Heng[2] said when he expressed his view on Lu Xun,[3] "The boundless gloom in Lu Xun's essays brightens up our own darkness."

One of the functions of art is to point out certain historical mistakes. Through realist films, we can face unpleasant facts in order to change and become happy and liberated people. What I said in Venice is still valid today: Filmmaking is liberating for me, and it could be a form of liberation for us Chinese people. What is the point of endless entertainment? Entertainment is harmless, and society should encourage various forms of entertainment.

Xu: You've said many times that you don't have a problem finding good scripts. Yet, the most important flaw of these Chinese blockbusters is weak storyline, despite their huge budgets. Writers recycle the plot of Cao Yu's *Thunderstorm* (*Lei yu*)[4] or even *Hamlet*. What are your thoughts on this?

---

2. Liu Heng, b. 1954, became a realist writer in the 1970s after working as a farmer, factory worker, and soldier.

3. Lu Xun (1881–1936) was a leading cultural and political Chinese leftist writer in his lifetime. He was a novelist, editor, essayist, translator, literary critic, poet, and political activist.

4. Cao Yu (1910–1996) was one of the most important Chinese playwrights of the twentieth century. *Thunderstorm* (*Lei yu*) is a classic Chinese family drama written in 1933. It has since been adapted into screenplays.

Jia: The common practice for screenwriters in the Chinese film industry is to capture the directors' thoughts and style. Screenwriters have to write well and help their directors express themselves—expressions of a real concept, a slice of life, or a new insight. If screenwriters cannot capture these, the script doesn't work, no matter how eloquent they are.

BECOMING THE CAPTIVE OF COMMERCIALISM

Xu: How did the blockbuster industry end up with this structure?

Jia: I went through the numbers and records a few years back. It seems that it got started when a few famous directors tried to change their styles and all happened to miss their marks. At the beginning of their transformations, they didn't open themselves to a broader artistic venture; they were just negating their earlier artistic values. They took a drastic plunge into a new genre, but their creativity was choked in terms of artistic and cinematic language. Their inner selves longed for renewal and transcendence, but they didn't have the energy or stamina to break through their bottleneck. So they turned to commercial cinema. Coincidentally, this was the time when many young directors became very active on the international stage.

Xu: Why did they all fall for commercialism?

Jia: From what I can gather, these directors are all a product of the times and prevailing trends. They weren't built for being innovative artists. The films and directors of *Yellow Earth* (*Huang tudi*) (1984) and *Red Sorghum* (*Hong gao liang*) (1987) are products of the times.

Xu: I believe you have high praise for *Yellow Earth*, right?

Jia: Indeed, I like it a lot. I'm not evaluating a specific person or his works, but I see this film in a social and historical context and recognize its effect on our culture. *Yellow Earth* belongs to the artistic or literary genre of root searching. The film embodies the great ideology of the 1980s. By the time of *Red Sorghum*, around 1987, reformation had run into obstacles, but the sense of strength, heroism, and elitism grew even stronger. The "Spirit of the Wine God" in *Red Sorghum* symbolizes this social trend.

If you look back at these films in context, they were adaptations of contemporary literature. That helped me understand them better. They form their narratives with the help of literature; they were a byproduct of the period in which literary adaptation was popular. I think these directors' independent thinking and judgment were limited.

Over the past twenty years, China has started to embrace a diversity of ideas and values. But creativity struggled. Directors failed to find values that the public cared about. As society fragmented, conflicting and paradoxical views of value proliferated. However, one value stood out at the time, and that was sales. Profitability became a god, and economic reforms reshaped society. Economic growth became the sole goal of everyone in the country, resulting in a nationwide economic obsession. Culture and ideology were completely marginalized.

Under these circumstances, producing commercial films with high investment and high return seemed to make sense. This gave them a good excuse to create a much-needed Chinese film industry. Their transformations were valid and reasonable.

So, people like Zhang Yimou and his producer Zhang Weiping turned into heroes defending the Chinese film industry against Hollywood. They claimed that if they weren't making blockbusters, then Hollywood would invade our market and that would be the end of Chinese cinema. I consider that declamatory. I don't believe that Chinese cinema would be destroyed without their films.

With commercial success, their egos became extremely inflated. But within their commercial operation, market success is not necessarily the ultimate goal. Their strategy is to partner with administration resources. If they didn't have the support of these resources, they would not be able to dominate the market. The whole nation's movie screens are so monotonous that it feels like we are back in the Cultural Revolution. But back then there were at least eight model operas and some small theater performances; now it's down to two or three blockbusters per year.

This policy of mobilizing all resources and favors is seriously hurting Chinese cinema. Most young directors have absolutely no way of screening their work in theaters.

BLOCKBUSTERS ARE LIKE A CONTAGION AND DESTROY SOCIAL VALUES

Xu: Even though independent directors do get a chance for theatrical release, the results are not at all satisfying. I read some statistics about how director Wang Chao's *Luxury Car* (*Jiang cheng xia ri*) (2006) only made RMB ¥402 [USD $66 in 2015 terms] at the box office in Nanjing.

Jia: Is that a fact? Since they control all the resources, they could manipulate the statistics to show that these small films are not profitable. We could take the release of *Hero* as an example. There was an administrative order that no Hollywood film was allowed to be released during the release of *Hero*—a clear indication of powerful administrative assistance.

Xu: So, since our theater chains are run by the state, the administration can directly interfere with theaters' commercial decisions?

Jia: Yes, they have common interests! The directors need to earn their reputation, and the producers need to make their money back, administrative departments need to break new box office records, and theater owners need to turn a profit. They cooperate to an unprecedented degree to reach their mutual goals.

Xu: Do you think these directors' changes were inevitable?

Jia: I do. As a director, if you don't own your heart and soul, you lose your individuality. They could not express themselves well in this age of disorienting diversity. The films they made that we admired were adaptations; they did not fully represent the directors' independent thinking. These directors borrowed elements of the rich cultural scene at the time and latched onto the prevailing ideologies in philosophy, literature, and art. So when we look at *Hero*, we see an admired director who stopped using literature as an entry point to commercial cinema. Dormant cultural conditioning surfaced in his work. He had experienced terrible oppression earlier in his life, so when the opportunity arose he jumped at the opportunity for power and accepted obedience as a virtue. The director who helmed *The Story of Qiu Ju* (*Qiu Ju da guan si*) (1992) became the defender of power in *Hero*.

I understand that nobody's cultural genetic composition is perfect. My point is that you should try to understand and overcome the flaws and accept new cultural views in order to conquer your own limitations. Established directors should not take criticism from their younger peers as "bad-mouthing." Zhang Yimou turned a casual discussion into a rant about how I was not being courteous and sincere. He was completely out of line, so I'm not going to respond any more.

Zhang Weiping responded to me with five points. I read them but have no interest in replying. I addressed him as a producer and a filmmaker, but he replied as a movie tycoon. He said that my attitude is that of someone that doesn't care for the rich. Well, of course I don't! When it comes to the art of film, I have nothing to share with tycoons. What good would it do to speak to him? I'm very disappointed by his response.

BLOCKBUSTERS ARE LIKE A CONTAGION AND DESTROY SOCIAL VALUES

Xu: Do you ever feel awkward about your published statements or when you are covered in the entertainment section?

Jia: That is why I'm talking to [you] today. I want to step away from entertainment and use another platform. I may not be able to reach a large audience, but this is not about numbers. There are certain cruel realities in our society; our discussion of cultural values should be frank. Maybe *Yellow Earth* never reached an audience of a hundred thousand, but the film had a very big impact on Chinese cinema and culture.

Xu: When you were speaking during the premiere of *Still Life* at Peking University, you said, "I'm curious, I want to find out who cares about the 'Good People' in this money-grubbing time." Was that intentional posturing or a genuine question?

Jia: I was indeed a bit puzzled. If *Still Life* were not premiering at the same time as *Curse of the Golden Flower*, I would have been shocked. *Still Life* and *Curse of the Golden Flower* are two very distinct films. They also represent two very different cultural attitudes. When the whole nation is obsessed with economic growth, how much space is left for developing culture in this atmosphere? How much have economic priorities eroded cultural sensibilities? My heart and my mind are not in sync on this issue. I had already drawn a rational conclusion, but the experience of this premiere really drove things home. That's why I have been more vocal and intense.

Xu: I understand that your persistence has convinced the producer to agree to premiere *Still Life* on the same day as *Curse of the Golden Flower*. Are you trying to give your rational judgment another chance?

Jia: I joked that this decision is my "performance art" piece. This action is definitely not a commercial decision, since that would be patently stupid. If we had selected another date to open, the box office return for *Still Life* would have been better. Since we set out to produce a work of culture, then we should insist on it till the end. We want to make a statement with action that there are still people who are not mercenaries today and filmmaking does not have to be based entirely on economic considerations. The way we release the film is an extension of its artistic expression. These days, when everyone is chasing profit, we want to leave behind a trace of culture.

Xu: And how did this release of *Still Life* turn out?

Jia: I think it was quite satisfactory; we have achieved all our goals. Our main concerns were addressed through the process of distribution. Public acceptance or box office results were less important. But looking at the audience numbers before New Year's Day, our box office return is already above RMB ¥2,000,000 [USD $320,000]. That's more than the RMB ¥1,200,000 [US $193,000] we got from *The World*.

Xu: So what you're saying is the release of *Still Life* shouldn't be interpreted as an economic move but a celebration of culture?

Jia: I have to point out to you that this is not our normal practice but a special occasion. The normal and healthy way to distribute a film should still be to put investors' interests first. This time we took aim at blockbusters—a Goliath—and we had no choice but to use unorthodox counter-actions. "Counter-actions" might be too soft; perhaps "attacks" is better.

Xu: You have called this an act of martyrdom for love. If you don't have the backing of overseas distribution, would you still be so bold to die for love?

Jia: If the overseas market didn't do well, or we hadn't recovered our costs, I may not have done the same thing. I'm not Superman; I can't afford to fall to pieces.

Xu: Can you tell us more about the release of *Still Life*?

Jia: The theater chains predetermined that the film would not draw crowds, so they assigned unfavorable screening times like 9am or 1pm, just to show you that you don't have an audience and that they could take it off the screens. The greed of the theater chains is undeniable, but it also demonstrated the irrationality of public operations. In France which is an even more commercially driven country, if a theater chain accepts a film for screenings, they give you a full week of screen time from 9 a.m. to 10 p.m.—five or six shows a day. After a week, they evaluate audience attendance. A week is the usual evaluation period. The theater owners assume the risk for a week once they commit to any film.

Things are different in China. They predetermine that you are not marketable and give you unfavorable screening times. They then tell you that your film is not marketable and pull the film off the screens. This practice is completely unfair. It is a rather dictatorial way of running a business.

Zhang Yimou once said angrily, "To each his own, don't you try to take screens away from us! You should build your own art-house cinemas!" I think he is right. We really should have different theater operations for different types of films. He was rather tactless, knowing that we only have one type of theater chain in China and it serves as a kind of public resource. We might need ten or twenty years to kick-start an art-house cinema operation. How can young directors wait that long? But you can also see that Zhang treats the nation's theater chain like his own property. To him, we were just interfering and being vexatious. Quite a boorish attitude.

Xu: Wang Shuo[5] once said, "We can't leave a legacy of films only about fighting for our future generations."

Jia: He is funny. But it is really a matter of our collective memory. Film should be a form of memory.

---

5. Wang Shuo, b. 1958, is a celebrated contemporary Chinese author, screenwriter, director, actor, and cultural icon. His works have been translated into other languages worldwide.

Entertainment is very popular and the public is bombarded by tabloid information, but insignificant memories should not be canonized. The most important things might be documented by arts that are not popular today. A positive society should help, encourage, and respect this type of work, not laugh at them like Zhang Weiping does. In their press release about us they wrote, "An insignificant film with a box office return of only RMB ¥200,000 [USD $32,000] trying to take on a RMB ¥200,000,000 [USD $32,000,000] production." You can read between the lines that they are denigrating artistic values, demonstrating the moribundity of this era, and exerting a terrible influence. I once heard a master's degree candidate at the Beijing Film Academy say, "Don't speak to me in terms of art!" They were ashamed of discussing artistry and I find that very sad. It seems like this attitude is even more in style today. It's really unfortunate that these values continue to spread.

AN UNAVOIDABLE TOPIC

Xu: Blockbusters are trying to say they are the only way to build an independent Chinese film industry. Do you think that is right?

Jia: I agree that we have to build our own film industry. That is totally justified. But these people are not really doing the work to achieve this goal. We should start with the basics to give young people an easier way to enter the film industry—to take down the barriers for them. Filmmaking and distribution should be separate from the power of government administration. They should operate independently in the market. There shouldn't be the present practice where government departments step in to help the promotion and sales of blockbusters.

Blockbusters should be made without a budget cap, but in a healthy film industry structure there should be enough room for mid- and small-budget films. The structure should resemble a pyramid, not be top heavy with no support at the middle. The reasonable way would be to allow a large number of film professionals to survive in the industry. Their experience and continued practice would ensure high standards for Chinese films. Blockbusters are all completed outside the country. They are finished in Australia or with post-production done in Hollywood. They do not contribute to the infrastructure of the Chinese film industry. They cast actors from Hong Kong, Taiwan, Korea, and Japan together with a few new Chinese faces. While they claim to be saving the Chinese film industry they are not really contributing much. What's real is the money they managed to extract from Chinese pockets.

Xu: Will you defy the opportunity to make a blockbuster yourself?

Jia: I really don't like this way of thinking. Budget should not be the main thing about a film. I'm planning to make a film in Shanghai about the revolution in 1927.

BLOCKBUSTERS ARE LIKE A CONTAGION AND DESTROY SOCIAL VALUES

The production will be expensive. I did not say that I wanted to make a blockbuster, so let's just do anything. That's putting the cart before the horse. Big-budget films shouldn't be their own genre. Budget figures should only be a means to an end for the film you want to make. Things should go the other way around.

Xu: What other things do you think should be corrected about today's film environment?

Jia: I think we should have a film rating system. *Curse of the Golden Flower* is listed as "not suitable for children" in the United States. Why are we encouraging our young people to see it here? The problem is that different societies have different standards.

Xu: As a director, you use films to narrate, document, and analyze your observations and express your views. What are your thoughts on this era?

Jia: I think there are some problems, like the turpitude of our youth culture and the youth's failure to rebel against this stronghold of morbid culture. They are ideologically knocked out by commercial values. Youth culture should always include rebellion to some extent, some ability to sense and resist unsettling issues. Like rock and roll and gospel music at the end of the last decade. I feel that youths should naturally lean toward these behaviors. They force society to evolve in a vivacious culture. Too bad that youth today only conform to commercialism. It's a real pity.

Universal commercialism in China today—not just blockbuster films—is transforming an entire generation of our youths. Speaking a bit more seriously, it's like a virus spreading in our community.

But there are also marks of progress. Unlike before, when materialism is allowed to succeed, it is definitely a sign of an open society. But material gain has turned into the only form of value for the people. People seem to think that if you don't seize more resources and get rich as fast as possible, your health and retirement expenses won't be guaranteed. The whole society just goes for it.

Xu: Do you think we could somehow avoid this stage?

Jia: Yes, by paying more attention to more than just commercial needs. Our current economic system is like robbery in broad daylight, and that's exactly what is being portrayed in our movies and reflected in the film industry, less so in other industries. I don't think real estate agents are as explicit about how much they invest and how much they get in return. Profit is a trade secret. No industry behaves like the film industry, where exposing profits is a means of generating more investments. For the same reason, everyone wants to earn praise and be a commercial hero.

Xu: It's strange that this way of living is right in front of our eyes, being repeated and continuing to succeed.

Jia: These changes are really quite recent. Our whole discussion just now has not been only about *Curse of the Golden Flower* and *Still Life* or the Chinese film industry. We are discussing Chinese culture writ large, our economic and political policies. It is an important topic and should not be avoided.

Originally published in *Chinese Youth Daily (Zhongguo qingnian bao), Ice Point Weekly (Bingdian zhoukan)*, January 10, 2007.

DECIPHERING CHINA THROUGH THE EYES OF A FILM-POET 40

A Conversation with Dudley Andrew, Ouyang Jianghe, Zhai Yongming, Lu Xinyu, and Jia Zhangke
Translated by Alice Shih

Ouyang Jianghe: We shall start tonight's discussion on Jia Zhangke's *24 City* with Professor Dudley Andrew from the Department of Comparative Literature at Yale University, where he is an expert on world cinema. He will speak to Jia first, followed by questions from the other participants.

Dudley Andrew: Jia Zhangke captivated me the moment I encountered his work. He is the first Chinese director I introduce to my students in America. My intuition was confirmed when I read his interview. I found out he was inspired after watching *Yellow Earth* by Chen Kaige, a film that also made a profound impression on me. I met Chen Kaige and Zhang Yimou at the Cannes Film Festival in 1985 and we watched the film together. I had never seen anything like it before. It was so mysterious and obscure, yet it was a very important ambassador for Chinese cinema on the world stage. At the time, young artists were fully aware that they had to take a new approach in presenting China to the rest of the world.

Like most film lovers, I'm very disappointed by the recent works of Chen Kaige and Zhang Yimou, although their influence remains ubiquitous and powerful. Their recent films are so different from Jia Zhangke's works. Jia is a poetic filmmaker; his films are about a nation and its people lost in the process of modernization. In an interview he gave about *Still Life* to the *Cahiers du Cinéma* in France, he was very open about his unique filmmaking ideas and attitudes. He talked about how his film was repressed by the *Curse of the Golden Flower* because of unfair distribution strategies. *Still Life* would be shown on the outskirts of Beijing, while the blockbusters, with their billboard marketing, dominated the theater chains in the capital and other major cities.

Interestingly, *24 City* has a particular green hue. I'm not sure if this tone is a projector effect or not, but I remember seeing the same color when the film was shown in New York. Is this the color of the Chinese spirit? I really like the hospital scene in *The World* where a relative comes to pay the peasant some money. The same green appears then, like the light emitted by fireflies.

Jia Zhangke: This green color is not a projector effect. We did that color timing in post-production, an achievement of digital technologies. In fact, we have used this color since the first night scene in our second film, *Platform*, where a few hundred peasants stand in front of a green wall waiting for a performance to begin. That same green filled the whole screen. That color always reminds me of the period between the seventies and the eighties, as most northern families in China painted their low walls green. As a small kid, I ran into this color daily not just around home but also at various institutions like hospitals, offices, schools, and all public places. At larger spaces like factory grounds, green continued to dominate, especially on state-owned machinery and walls. I'm very familiar with that color and it matches my memory of the last decade under the old system.

Andrew: My ancestors are Irish and I really like Yeats, whose lines you incorporated into your script. It reads as follows:

> We that have done and thought,
> That have thought and done,
> Must ramble, and thin out
> Like milk spilt on a stone.[1]

I'd like to ask Jia and his screenwriter, Zhai Yongming, why they chose that quote. Were you trying to guide the audience somewhere specific when those lines appear between scenes?

Jia: Zhai and I were discussing the script before the shoot and decided we wanted to incorporate a lot of poems and lyrics to express memories and stretch out the sense of time passing. When I was new to filmmaking, I had strong feelings about the fast pace of most contemporary films; they are all about action. But I think that human beings possess a lot of complicated emotions that spoken or written words can better express. So what was stopping us from just going

---

1. "Spilt Milk" was first published in *The Winding Stair* (1933).

back to words? We could use interview footage and the subjects' descriptions of their lives. So I selected a line from Ouyang Jianghe's book, *Glass Factory* (*Boli Gongchang*), which says, "The whole aviation manufacturing plant takes on the form of a huge eyeball/ and manual labor is the blackest part in the middle," as well as another line from a poem by Wan Xia.[2] The quote from Yeats was selected by Zhai for me and I was very excited about it.

Zhai Yongming: William Butler Yeats is my favorite poet. No one has inspired me more, and I haven't stopped reading his poems. When we were writing the script, Jia spoke of a shot where there is a blast at the factory followed by dense smoke and the arrival of the youngest characters—the latest generation. I find this shot very poetic, and even though I didn't have a chance to watch the shoots at the time, the Yeats quote had already occurred to me and I couldn't get it out of my mind the moment I heard Jia's description.

I see Jia as the most poetical and lyrical of all Chinese directors. He has studied poetry in depth. Many of the lines of poetry in the film were written by him. I was responsible for the other selections. Therefore, I think what Professor Andrew said earlier about Jia being a poetic filmmaker is very appropriate.

Andrew: Jia is indeed a poet, and he cares about society very deeply. On another topic, why was it necessary to select nine characters for the script of *24 City*?

---

2. Wan Xia is a contemporary poet from Sichuan born in 1962.

Jia: These nine characters are grouped into two categories. The first are actual people I interviewed inside the factory. We contacted more than a hundred people and shot about fifty of them. Out of these, I selected five subjects to put in the film. The other four characters are fictitious: Lü Liping's story about losing her child, Joan Chen's story of a woman in Shanghai at the end of 1970, Chen Jianbin's childhood story during the Cultural Revolution, and the story of the new generation of people portrayed by Zhao Tao.

These nine characters come together in my mind to form a complete picture. On one hand, I really like the feeling of portraying a story following only one character and the people around this person, but I didn't want to use just one fixed character throughout the whole shoot. So I chose the concept of a "group portrait" and brought many characters together in this film. These nine characters interact with each other in a chronological order from 1950 to the present in a relay, narrating history in a linear progression. Each character inhabits his or her unique time period.

All nine of them are narrating right in the present moment, but they could be talking about anything that happened in the past fifty years. I like this temporal complexity. During the fifty-odd interviews we shot, there were some heated discussions and frightful moments, but I decided that should be edited out. What remains is just the narration of some common experiences. For most Chinese people, these daily life experiences are very communal, not pertaining to any special individual. These shared experiences could lead the audience to a greater imaginative space where they include themselves in the situation. This film is not about a specific case; it is a collective memory.

There are a lot of quiet moments in the film, such as portrait shots with no dialogue. Everyone goes through such quiet moments, and they are complementary to spoken words.

Andrew: Could *24 City* be a mirror of your film *The World*? The people in *The World* enjoy unrestricted freedom while the people in *24 City* are not allowed to travel as they wish. Yet inside this walled society, they interact, help each other, and build powerful memories together. But the people in *The World* are all lonely and homeless. Does that have to do with China's modernization?

Jia: That has a lot to do with the period before we shot *The World*. The script of this film was written in 2003, and it has a lot to do with SARS. Before SARS, I was like most Chinese people, living life in the fast lane. The arrival of SARS put a brake on everything, and I was trapped in Beijing but couldn't get anything done. When I wandered around the city, I suddenly understood the problem with fast-paced living. But we didn't notice how fast we were going until the brakes were pulled.

If you asked me about the main thing I learned through making this film, it would be a new critical attitude of the fast-paced life of the city. The appeal of speed is a new concept. I noticed the advertisements along both sides of the streets as I walked, and I saw some new housing developments called "Roman

Garden," "Vancouver Forest," or "The Water City of Venice." All these names of new developments in Beijing are associated with foreign cities. I found this a bit unsettling. It may be affecting our minds and spirits.

The World and *24 City* are indeed different. *The World* deals with people who enter a new city, liberated from their families, to start a new life. By relocating, they are, in fact, leaving two things behind. One is their hometown where all their relatives reside and the other is the family structure that has always been the core of Chinese society. These people are far away from their families, so family members and their values no longer have a strong hold on them. The also walk away from their previous work units in factories and so on. The living quarters and the factory in *24 City* are like the destinations for the relocated people in *The World*. "Work units" are the main way the Chinese population has been structured since 1949. All the people in *24 City* fall into this social structure. I entered the factory and was surprised to find the people inside were very detached from the wider world. Although this factory is part of contemporary China, it feels very disconnected from modern-day China. Outside the wall of the worker quarters, there are modern establishments like travel agencies, shops, nightclubs, Internet cafés, bars, and everything else. Inside their units, I saw furniture, decor, bathrooms, photos, and displays stuck in the late 1970s and '80s. It's worlds apart inside and outside those walls.

I think I'm doing two things through these two films. *The World* shows a new China being reinvented through progress, while *24 City* shows a China locked in the past.

Andrew: *Platform* ends on a tone of pessimism toward self-expression in China, but this is an area in which we have made considerable progress. How do you explain all this development and change in Chinese self-expression?

Jia: I honestly don't feel that this sense of disappointment came as a simple reflection of events in the real world, in the way that one's past creates tragic disappointments later. According to Buddhism, the progression of life has four stages: birth, age, sickness, and death. These four stages will not fade or disappear, regardless of any increases in our ability to express them or the pressures of society. From this perspective, my understanding of life is not that happy.

Expression evolves differently in every era. As I look back on the period between the late seventies, when the nation started reforming its policies, up till now, I think it has been important to learn an openness to express our personal points of view and not to be attached to mainstream attitudes. There should be diverse means of expression, and I think some progress is being made on this. More artwork is being made from a personal viewpoint in China today, especially in the medium of film. Underground independent films that started in the nineties are now being recognized. Some of them offer unique observational views on society and reflect the lives of the people from a personal angle. We should appreciate this big change in the past ten-odd years. We know that Super 8 films were used in the sixties and seventies to record personal memories in Japan. When we look back at the important historical moments in China this century, we only find state-recorded images. We have none from the point of view of regular citizens. Finally, in the nineties, China saw the arrival of independent films and free expression. For the past eighteen years, I believe we have finally generated some meaningful works that reflect personal views. Our self-expression is maturing.

Lu Xinyu: I find it difficult to discuss things with Director Jia, because Jia leaves very little space for his critics. His films speak best for themselves, as the views and messages in his films are structured very tightly.

My question about *24 City* concerns the interplay between film forms, how it breaks traditional film narrative styles. So, deliberately switching documentary forms, making the director take the role of interviewer, switching shots from inside to outside. This crossing-over process brings out a contradiction. When you place yourself inside the film, you seem to be limiting your influence. You show what you capture on film through the interviews, but you do not divulge things that were not recorded on camera. The paradox is that what you did not capture on camera might be the most important material. You have chosen the traditional documentary that only shows what the camera sees, relinquishing the all-knowing form of narration with the omnipresent camera. You come back to reality to tell your story. However, deliberately adopting this limited angle of expression conflicts with the view of a poet. The appearance of lyrics and songs in the film seem to be the words of a prophet or a judge. That is, based on your omniscient and transcendent perspective, you offer us this prophetic and judging knowledge.

The whole story is about losses in the past: a kid was lost, love lost, youth, years, etc. Everything was lost in the past. Since they were lost, you could not show them on screen. Therefore you appear to be restrained or even trapped at times. This form of mockumentary, with a mock interviewer, is in fact a form of "anti-narrative" narrative film.

There is another minor detail: most of the people in the film speak their own dialect; even you were speaking with an accent. Why would you let Lü Liping speak in standard Mandarin? I felt very strange and uncomfortable when I heard her voice. It was like the film stopped and the voice track turned into lines from a television drama. This seems to be the only flaw in the film.

Jia: When I was in the process of designing this film, I was hoping the film form would be a kind of hybrid. Since there is continuity between the temporal and visual elements in films, there is a possibility for some cross-media experiment. The many interviews borrow from oral history; the verses are borrowed from literature. There are also borrowed images and music. My first idea was to mix media, hoping for a complete, multilayered work built through a mixture of art forms. When I get in touch with memory, I find it very complex, and trying to reveal it calls for an intricate process. When I was doing the interviews, the relationships between the director and the subjects, or between the director and the film, all felt complicated. At the end, I thought I should put all these convoluted elements and their corresponding complexities together. The character of Lü Liping actually migrated to the factory from Shenyang. Mandarin and the Northeastern dialect are similar, and Lu's husband is from the Northeast. I thought she could speak some Northeastern dialect but she was unable to. She doesn't even carry a trace of Northeastern accent, so that's definitely our loss not to be able to express the dialect. However, Lu does have a big advantage, and that's her ability and control over her delivery. It was impossible to locate this original narrator, as she is already dead. This story is about the first child who died at the factory. I thought Lu was appropriate because she is a mother who lived as a single mom for a long time. Her relationship with her child, and her feelings as a mom, come out powerfully as a result of her personal experience. I saw that she really understood the essence of the script, better than other actors. So I accepted the sacrifice of the dialect.

As for the method of crossing media, I was hoping for more freedom. I hoped to get in touch with the vibrancy of early silent cinema, where there were no cities and no delineation between documentaries and narratives. Films were just images recorded on celluloid. Since no one could actually figure out what the medium of film was, there was a lot of freedom at that time. Silent films have captions, and they could be very literary. I was trying to review the state of cinema today. Whether narrative or documentary, I think all films have fallen into limitation and I am looking for a way to break this spell. I realize this method will not make it into the mainstream. Like you said before, this new film form does not break all the traditions in Chinese cinema. But perhaps it broke some; and perhaps the effort will lead to other possibilities. But in general it was a tribute to the silent era.

Ouyang: Maybe all poets would enjoy working the same way as Jia. His works exhibit images and memories. The way he interprets the world is just like what Professor Andrew poignantly described earlier. Jia could be described as a poet who uses images to construct his poems. Poems cross different forms too. Apart from words, they employ images and sounds. As Zhai Yongming said, she was hoping to go beyond the scope of words. By incorporating lyrics into the script, she ventured into the aesthetics of music, visual art, and films. She knows a lot about filmmaking, so I consider this collaboration with Jia very successful.

Jia mentioned that he shot fifty interviews and chose to take out all the over-the-top and disturbing footage. Yu Jian[3] was a worker before and found the

---

3. Yu Jian, b. 1954, is a contemporary Chinese poet. He quit school at the age of fourteen, began writing poems at twenty, and got published at twenty-five after being a worker at various trades. He is currently teaching at Yunnan Normal University.

lives of these workers too straightforward and not exciting enough. But the simplicity made the film more abstract and lifted their experience to a higher level, a somewhat fabricated height. The poem at the end shatters all of our conceptions of film; it is not a linear story, narrative, documentary, television drama, or commentary. It is none of those but expresses all of them. It's very strange—this kind of self-contradictory, mutually demeaning thing, compressed into fanciness, the lower common denominator of entertainment, and now poetry appears. But its purpose isn't to elevate the film; it's a compressive, reductive, subtractive process. Boiled down to nothing, it becomes a bare, dried-out thing: our historical text, our memory.

I really admire Professor Andrew's insight about the green color in Jia's film. The Chinese painter Zhang Xiaogang[4] also remarked on this. His retrospective of the seventies also features green heavily. In his essay "Less than One," about his early life, the Russian American poet Joseph Brodsky talked about a line running along the walls in the Soviet Union, post offices and so on, painted green—I'll call it "postal green"—up to 1.2 meters [4 feet]. In China the standard was 1.1 meters [3.5 feet], and there was a standard color and shade. This "postal green" is very significant: it's a product of socialism and not to be found anywhere in capitalist countries. Why this is the case is a mystery.

Professor Lu spoke of *24 City* as a film about losses in the past. These losses are not just figurative—there have been concrete ones as well. For example,

---

4. Zhang Xiaogang, b. 1958, is a contemporary Chinese painter.

the financial crisis with the fall of Lehman Brothers destroyed many people's entire life's savings. This story of relocation also relates to the loss of the factory. There will be a new kind of city with the rise of private housing developments. Living in these privately invested, non-public housing developments, residents may never connect very intimately with the people around them. All they have in common is where they live, but their jobs, passions, educations, and cultural backgrounds are all different. All they need is money to move in there. The space of those who have moved here becomes an independent colony. They come from Shanghai, Liaoning, and other places and become a cultural exclave, completely disconnected from the city around them. They left their hometowns in the 1950s to build their own ideal society here. They uprooted and have no more hometowns to go back to. Fifty years later, everything is gone. This is very meaningful.

Originally published in *21st Century Business Herald* (*21 Shiji Jingji Baodao*), November 8, 2008.

INDEX OF NAMES

Alain Resnais 191
André Bazin 191
Andrei Tarkovsky 139
Andrew Cheng Yusu 206
Ang Lee 186-188
Asako Fujioka 177, 182
Bai Guang 164
Bei Dao 78, 138
Cao Fei 201
Cao Yu 303
Chen Danqing iii, 19
Chen Jianbin 323
Chen Kaige 23-24, 75, 118, 205-206, 319
Chen Yingzhen 156
Chen Yusu 203
Chow Keung 225
Chow Yun-fat 38, 186
Donald Richie 211
Dudley Andrew vi, 319
Du Haibin 199
Dustin Hoffman 192
Edward Yang 183-184, 187-188
Emperor Qianlong 286
Étienne Chatiliez 77
Fei Mu 2, 8, 205
Fellini 159, 195-196, 283
Feng Xiaogang 23
Francis Ford Coppola 206
F. W. Murnau 138
Gan Xiao'er 206
Gong Xue 125
Grace Chang 164
Gu Zheng 131-133

He Jianjun 76, 197
Hirokazu Koreeda 179
Hiroshi Yanai 178
Hou Hsiao-hsien iv, 23, 31, 88, 153, 162, 165, 177, 195, 210, 214, 219
Hu Shu 199, 201
Ichikawa Shozo 273
Jean-Luc Godard 24, 139
Jean-Pierre Léaud 171
Jeff Chang 88
Jian Ding 87
Jia Zhangke iii, iv, v, vi, 1-4, 6, 9, 11, 17, 19-24, 26, 51, 63, 89, 123, 153, 161-162, 165, 201, 203, 229-230, 243, 247, 297, 319-320
Jim Jarmusch 3
Jin Shangzhe 206
Joan Chen 71, 323
Joel Coen 206
John Akomfrah 149, 166
John Woo 38, 118, 194
Joseph Brodsky 331
Juliette Binoche 194
Keiko Araki 177, 180
Keisuke Kinoshita 213
Ken Jones 225, 227
Kim Ki-duk 274
King Hu 126, 186
Kitano Takeshi 273
Kogo Noda 214
Kōhei Oguri 258
Krzysztof Kieslowski 145, 258

Kurosawa 132, 195, 209, 258
Lee Kang-sheng 171
Liang Jingdong 112
Li Baotian 286
Lin Xiaoling 2, 33, 133, 143-144
Lin Xudong iv, 19-21, 23, 89, 182
Li Tuo 191
Liu Heng 303
Liu Xiaodong v, 21-23, 67-68, 167-168, 229-231, 237, 239, 243, 245-247, 252-253, 289-290
Lou Ye 197
Lu Chuan 206
Lü Liping 71, 323, 328-329
Lucian Freud 237
Luis Buñuel 196
Lumière brothers 141
Lu Xinyu vi, 319, 327
Lu Xun 188, 303
Lu Yao 163
Ma Feng 152
Maggie Cheung 185
Ma Ke 69, 167, 247, 249, 251-252
Marcel Duchamp 82
Martin Scorsese v, 3, 223-226
Matthieu Laclau 6
Meryl Streep 192
Mian Mian 203
Michelangelo Antonioni 52, 191
Mikio Naruse 213
Mori Masayuki 273
Naomi Kawase 179, 182
Orson Welles 261
Ouyang Jianghe vi, 319, 322
Philip Kuhn 202
Polanski 260
Prime Minister Liu Luo Guo 286
Proust 193

Quentin Tarantino 260, 263
Rainer Werner Fassbinder 243
Richard Gere 195
Robert Bresson 102, 268, 278
Rohmer 195, 260
Romain Rolland 286
Sakyamuni 80
Seijun Suzuki 243
Sergei Eisenstein 139
Shen Congwen 9, 201
Shinsuke Ogawa 260
Shozo Ichiyama 177, 215
Song Yongping 148
Steven Spielberg 225
Su Li 152
Sun Jianmin iv, 123
Sylvia Chang 166
Takenori Sento 182
Teresa Teng 42, 292
Teruyo Nogami 132
Tian Zhuangzhuang 205
Tony Leung Chiu-wai 185
Tony Rayns v, 243, 247, 257, 274
Truffaut 24, 171, 223, 262
Tsai Ming-liang iv, 149, 161, 165, 274
Tsui Hark 194
Ulrich Gregor 119
Vittorio De Sica 102
Wang Bing 161, 199-200, 206
Wang Bo 133
Wang Chao 307
Wang Fen 199-200
Wang Hongwei 2, 105, 108-110, 131-132, 142, 173, 226
Wang Meng 87
Wang Mo-lin 171
Wang Shuo 312

INDEX OF NAMES

Wang Weiguo   163
Wang Xiaoshuai   20, 76, 197
Wan Xia   322
Wei Tie   11
Werner Herzog   77
William Butler Yeats   322
Wim Wenders   210, 264
Winston Chao   285
Wong Ain-ling   258
Wong Kar-wai   185-188, 222, 243
Wu Cheng'en   81
Wu Tianming   163
Wu Yonggang   10
Xi Chuan   76, 148
Xie Jin   25
Xu Baike   vi, 297
Xu Bing   11-12
Xu Yuan   196
Yang Fudong   201-202
Yang Tianyi   199
Yang Zaibao   283
Yasujirō Ozu   8, 177, 209-210, 215, 219
Ying Weiwei   199-200
Yōichi Sai   182
Yuan Muzhi   8, 218
Yu Jian   330
Yu Lik-wai   86, 98, 114-115, 142, 185, 223
Zhai Yongming   vi, 319, 321-322, 330
Zhang Ailing   9
Zhang Ming   76, 202
Zhang Nuanxin   191
Zhang Weiping   306, 308, 313
Zhang Xiaogang   331
Zhang Yangqian   144

Zhang Yimou   13, 23, 75, 119, 144, 162, 205-206, 238, 301, 306, 308, 312, 319
Zhang Yuan   20, 76, 197, 202
Zhang Ziyi   186
Zhao Dan   219
Zhao Peng   133
Zhao Shuli   152
Zhao Tao   3, 16, 71, 173, 183-185, 293, 323
Zhao Xuan   219
Zhong Hua   201
Zhu Wen   202

INDEX OF FILMS & MEDIA

"A Filmmaker inside the System." 206
"Answer" *(Huida)* 138
"*In Public* in My Own Words" 5, 51-56
"Less than One" 331
"On the Modernization of Film Language" 191
"Spilt Milk" 321
"The Age of Amateur Cinema Is about to Return." 55, 131
21st Century Business Herald (21 Shiji Jingji Baodao) 332
24 City iv, 3-4, 71, 319-320, 322, 324-327, 331
3-Iron (Bin-jip) 274
A Better Tomorrow 194
A Brighter Summer Day 183
A City of Sadness (Bei qing cheng shi) 153
A Conversation with God 165-166
A Straightforward Boy 212
A Touch of Sin 2, 4-7, 12-15
A Touch of Zen 186
A Tribe of Generals (Jiangjun zu) 156
Along the Railway (Tielu yanxian) 199
An Estranged Paradise (Mosheng tiantang) 202
Art World (Yishu shijie) 56
Avant-garde Today (Jinri xianfeng) 34, 47, 123, 149

Battleship Potemkin 194, 228, 261-262
Beijing Bastards (Beijing za zhong) 76, 197
Beijing Evening News 2, 189
Beijing Youth Daily 33
Beijing Youth Weekly (Beijing qingnian zhoukan) 196
Bell Flowers (Lingdang hua) 156
Bicycle Thieves (Ladri di biciclette) 159
Black Breakfast 14
Blood Is Always Hot (Xue zong shi re de) 283
Blue House (Lan fang zi) 78
Bonnie and Clyde 139
Book from the Sky 12
Breaking the Waves 264
Breathless 24, 190, 260
Buena Vista Social Club 264
Café Lumière (Kôhî jikô) 157
Cahiers du Cinéma 203, 320
Chain (Lian) 201
Chengdu Economic Daily (Chengdu shangbao) 152
China Pictorial 115
Chinese Cinema 2006 (Hua yu dian ying 2006) 284
Chinese Youth Daily (Zhongguo qingnian bao) 297, 317
Cinema Paradiso 123
Citizen Kane 194, 261
Classic of Poetry (Shijing) 158

Crouching Tiger, Hidden Dragon (Wohu canglong) 186-187
Cruel Story of Youth 24
Curse of the Golden Flower (Man cheng jin dai huang jin jia) 13
Dacheng Newspaper (Dachengbao) 286
Dogme 95 Manifesto 264
Dong iii, 21, 67, 76, 153, 167-168, 201, 229, 232, 234, 244, 247-248, 289
Dragon Gate Inn 126
Dudu 33, 132-133, 143
Elite Reader (Chengpin haodu) 160, 173
Endless Love (San bat liu ching) 221
Equinox Flower 213
Even Dwarfs Started Small 77
Farewell My Concubine 112-113, 118
Film Comment 274
Flight of the Red Balloon (Le voyage du ballon rouge) 157
Flowers of Shanghai (Hai shang hua) 88, 155
Forrest Gump 189
Gangs of New York 225, 228
Glass Factory (Boli Gongchang) 322
Good Morning, Beijing (Beijing ni zao) 221
Great Immigration at the Three Gorges (Sanxia da yi min) 229, 234
Guerrillas on the Plain (Pingyuan youjidui) 220
Guide to Quality Shopping (Jingpin gouwu zhinan) 196

Help Me Eros 171
Hero (Ying xiong) 301
Hey, Sun Is Rising! (Hei, tian liang le) 202
House of Flying Daggers (Shi mian mai fu) 301
I Was Born, But... 212
I Wish I Knew 4, 6
Ice Point Weekly (Bingdian zhoukan) 297, 317
In Public (Gonggong Changsuo) 5, 49-56, 149, 166-167, 201
In the Mood for Love (Fa yeung nin wa) 222
It's Real Cold Here (Zan zher zhen leng) 206
Jia Xiang 2-4
Jia Xiang 2 4
Journey to the West (Xiyouji) 80
Kinema Junpo 211, 213
Kramer vs. Kramer 192
L'Age d'Or 260
La Strada 159
La vie est un long fleuve tranquille 77
Last Year in Marienbad 191
Late Spring 210
Leave Me Alone (Wo bu yao ni guan) 201
Legends of the Fall 114
Libération 183
Life (Ren sheng) 163
LIFE Magazine 228
Luxury Car (Jiang cheng xia ri) 307
Maborosi 179
Mice Love Big Rice (Lao shu ai da mi) 292

INDEX OF FILMS & MEDIA

Millennium Mambo (Qian xi man po) 157
More than One Is Unhappy (Bu kuaile buzhi yige) 200
Mountain Path (Shanlu) 156
News and Newspaper Summary 126
Nights of Cabiria 191
Not One Less 144
Old Men (Lao tou) 199
Once Upon a Time in China 194
One Day, in Beijing (You yitian, zai Beijing) 33, 55
One Hundred Years of Solitude 191
Oriental Art (Dongfang Yishu) 241
Our Youngsters (Wo men cun li de nian qing ren) 152
Pia 177-182, 342
Platform (Zhantai) 273
Popular Movies (Dazhong dianying) 125
Postman (Youchai) 76, 197
Railroad Guerrilla (Tiedao youjidui) 220
Rainclouds over Wushan (Wushan yunyu) 76, 202
Raining in the Mountain 126, 186
Rebels of the Neon God 164, 172
Red Corner 195
Red Desert 191
Red Sorghum (Hong gao liang) 305
Remembrance of Things Past 193
Rice (Da hong mi dian) 207
Satellite TV Weekly (Weishi zhoukan) 83, 88
Seafood (Haixian) 202
Seeking Advice (Wen dao) 11
Seventeen Years (Guo nian huijia) 202
Shanxi Literature 93
Shanxi Youth 93
Sight & Sound 257
Soho Post 287
Sons (Erzi) 197
Soulstealers: The Chinese Sorcery Scare of 1768 202
Southern Metropolitan Weekly (Nanduzhoukan) 275
Southern Weekly (Nanfang zhoumo) 182, 188, 260
Spring in a Small Town (Xiao cheng zhi chun) 205
Still Life (Sanxia Haoren) 229
Storm (Fengbao) 25
Street Angel (Malu tianshi) 8, 218
Suzaku 179
Swan Song (Juexiang) 221
Sword of Penitence 212
Tender Is the Night 98
The 400 Blows 24, 123, 191
The Age of Innocence 227
the Beijing News (Xin jing bao) 222
The Box (Hezi) 200
The Boys from Fengkuei (Feng gui lai de ren) 155
The Condition of Dogs 14-15
The Days (Dongchun de rizi) 76, 197
The Deer Hunter 192
The Goddess (Shennü) 10
The Godfather 19, 139, 190, 194, 262
The Graduate 194
The Killer 38, 118, 194
The King of Comedy 223
The Lion King 184
The Metamorphosis 191

The Mirror 190
The Missing Gun (Xun qiang) 206
The New Immigrants of the Three Gorges (Sanxia xin yi min) 229
The Only Son 213
The Only Sons (Shan qing shui xiu) 206
The Puppetmaster (Xi meng ren sheng) 153
The Seagull (Sha ou) 191
The Shawshank Redemption 189
The Story of Qiu Ju (Qiu Ju da guan si) 308
The Wayward Cloud 172
The White-haired Girl (Bai mao nu) 156
The Winding Stair 321
The World (Shijie) 59-64, 236, 248, 292, 310, 320, 324-325
Third Edition of the Chinese Film Media Awards (Huayu dianying chuanmei dajiang) 207
This Winter (Jin nian dong tian) 201
Those Chinese Republicans (Minguo Naxieren) 297
Three Colors, Blue 194
Early Spring (Zao Chun Er Yue) 25, 214
Thunderstorm (Lei yu) 303
Today (Jintian) 121
Together (He ni zai yi qi) 205
Tokyo Story 210-211
True Lies 194
Two Butterflies (Liang zhi hu die) 292

Un chien andalou 201
Unknown Pleasures (Ren xiao yao) 13, 57-58, 60, 171, 224-225
Useless iv, v, 69, 103, 167, 247-249, 251-252
Vive L'Amour (Ai qing wan sui) 274
Warm Bed (Wenchuang) 67, 229, 233
Water Margin (Shui Hu Zhua) 7, 81
We're Scared (Women Haipa) 203
Weekend Lover (Zhoumo qingren) 197
Welcome to Destination Shanghai (Mudidi Shanghai) 206
West of the Tracks (Tiexi qu) 200, 206
World Screen (Huanqiu yinmu) 215
Xiao Shan Goes Home (Xiao Shan Hui Jia) 29-34, 98, 110, 132-133, 141-143
Xiao Wu (Pickpocket) 35-40, 82, 85-86, 89, 92, 98, 100-102, 104, 107, 109-112, 114, 116-118, 123, 132-133, 138, 141-143, 146-147, 152, 154, 160, 171, 173, 203, 224-227, 259, 286
Yellow Earth (Huang tudi) 24, 206, 305
Yi Yi (A One and a Two) 183

INDEX–OTHER

"Reform and Opening Up" policy 190
Baoji 199
Beidaihe 202
Beihang University 143
Beijing Film Academy 2, 8, 85, 96, 123, 131, 189, 197, 199, 217, 223, 269, 313
Beijing University of Aeronautics and Astronautics (Beijing hangkong hangtian daxue) 86
Beijing University of Chemical Technology 87
Beitaipingzhuang 180
Berlin Film Festival 119
Big Bell Temple 261
Box Bar (Hezi) 196
Boxer Rebellion 4
Broadcasting Institute 19
Busan International Film Festival 185, 258-259, 273
Cannes Film Festival 2, 6, 153, 179, 224, 319
Central Academy of Fine Arts 19-21, 89
Chang'an Avenue 138
Changchun 151, 196
Changchun Film Studio 151
Changchun Film Study Group (Changchun dianying xuexi xiaozu) 196
Chengdu 71, 152, 337
China Central Television 33, 301
Dai Luoding Temple 278

Datong 15, 51, 57-58, 149, 281-282
Drama Troupe of the People's Liberation Army's General Political Department 199
Dubao/Derby cigarette 86
Eighth Route Army 24
Engaku-ji monastery 177, 215
Fengjie 65-68, 229-230, 290
Free Film (Ziyou dianying) 196
Fubon Charity Foundation 165
Fujian 193
Fukagawa District 210
Fukuoka Art Museum 229
Great Cultural Revolution 22, 25
Great Hall of the People 282
Great Western Development policy 51
Guangdong 193
Guangzhou 42, 45, 69, 196, 201, 206, 249, 251
Guangzhou Academy of Fine Arts 201
Guizhou 199, 201
Haidian District 86
Hong Kong International Film Festival 89, 258
Hongkou Cultural Palace 196
Hôtel de Ville 186
Hu Tong Communication 98
Huang Ting Zi (the Yellow Pavilion) 85
Huang Ting Zi Number 50 85
Imperial Hotel 181

International Documentary Film Festival of Marseille 161, 201
Japanese "I-novels" (watakushi shosetsu) 200
Jeonju International Film Festival's Digital Short Films by Three Filmmakers program 51
Jiangxi 19, 199-200
jiazi 209
Jinan 196
Jinci Temple 93
Jinxian 200
Joint Publishing bookstore 75
Kamakura 177, 210, 215
Kuomintang's White Terror 156
Le Centre Pompidou 183
Lehman Brothers 332
Liaoning 332
Loess Plateau 96, 125
Madian 85
Militia United in Righteousness 4
MK2 194
Mount Wutai 277
Museum of Contemporary Art Shanghai 229
Nanjing 196-197, 202, 307
National Palace Museum 168
New York Film Festival 225
North East 20, 99
Office 101 (101 Bangongshi) 195
Office Kitano 273
People's Cinema (Pingming dianying) 196
Fanhall Films (Xianxiang gongzuoshi) 196
PIA Film Festival 177-182
Pingyao 125
Practice Society's Yellow Pavilion Images Online (Huang tingzi yingxiang) 197
Qixian 125
Queensland Art Gallery 229
Rear Window Film Viewings (Houchuang kan dianying) 196
San Francisco International Film Festival 225
San Francisco Museum of Modern Art 229
Sanxia iii, 65, 153-154, 229, 234, 289-290, 294, 338
Seven Sages of the Bamboo Grove (Zhulin qi xian) 82
Shaanxi 125, 163, 199
Shanghai Drama School 123
Shanghai University 203
Shanxi 14-15, 20, 35, 51, 60, 69, 82, 86, 89, 93-96, 123, 125, 129, 133, 135, 149, 152-153, 248-249, 251, 277
Shanxi University 82, 93-94
Shanxi Writers' Association 93
Shaoshan 124
Shenyang 71, 196, 199-200, 206, 329
Shochiku Film Company 211
Sichuan 86, 99, 199, 245, 249, 293-294, 322
Sina Weibo 5
Southern Film Forum (Nanfang dianying luntan) 196
Taipingzhuang 85
Taiyuan 51, 65, 82, 93-95, 278
the Bund 219

the National Art Museum of China 229
Three Gorges Dam 65-67, 230
Tiananmen Square 33, 247
Tiexi 200, 206, 339
Toho 179
Tokyo International Film Festival 259, 273
Toronto International Film Festival 16, 183
Tsinghua University 196
Venice International Film Festival 153, 161, 183
West Tayuan 121
Wuhan 196
Wuhan Film Watchers (Wuhan guan ying) 196
Xi'an Film College 10
xiansheng 11
Xiaoxitian District 267
Xiaoyi 91, 127
Xin Ma Tai 85
Xinghuacun Fen Wine Factory 127
Xinjiang 51, 86
Xinjiekou 85
Xuxi 94
Yamagata International Documentary Film Festival
Yan'an 25
Yanfen Street 200
Yellow River 24, 96, 125
Youth Experimental Film Group 2, 34, 130-131
Youth Palace Cinema 205
Yulin 279
Yunnan Normal University 330
Zhuhai 247, 252

loafer
26 "sent-down youth"